Melanin, the Master Molecule

Edited by

Arturo Solís Herrera
Director and Founder Human Photosynthesis® Study Center Aguascalientes México

General:

1. Any dispute or claim arising out of or in connection with this License Agreement or the Work (including non-contractual disputes or claims) will be governed by and construed in accordance with the laws of the U.A.E. as applied in the Emirate of Dubai. Each party agrees that the courts of the Emirate of Dubai shall have exclusive jurisdiction to settle any dispute or claim arising out of or in connection with this License Agreement or the Work (including non-contractual disputes or claims).
2. Your rights under this License Agreement will automatically terminate without notice and without the need for a court order if at any point you breach any terms of this License Agreement. In no event will any delay or failure by Bentham Science Publishers in enforcing your compliance with this License Agreement constitute a waiver of any of its rights.
3. You acknowledge that you have read this License Agreement, and agree to be bound by its terms and conditions. To the extent that any other terms and conditions presented on any website of Bentham Science Publishers conflict with, or are inconsistent with, the terms and conditions set out in this License Agreement, you acknowledge that the terms and conditions set out in this License Agreement shall prevail.

Bentham Science Publishers Ltd.
Executive Suite Y - 2
PO Box 7917, Saif Zone
Sharjah, U.A.E.
Email: subscriptions@benthamscience.org

BENTHAM SCIENCE

CONTENTS

PREFACE i

DEDICATION ii

INTRODUCTION iii
PRESENTATION iii
Melanin Functions in the Eye iv
The Omnipresence of Melanin in All Biological Realms iv
Melanin, the First Self-replicating Molecule lxv
Melanin, the Molecule Base of Biological and Planetary Systems lxvii
CONCLUSION lxvii
REFERENCES lxix

CHAPTER 1 INTRINSIC CHEMISTRY OF MELANIN: POTENTIAL APPROACHES FOR TRANSLATIONAL MEDICINE 1
Arturo Solís-Herrera and *Sergey Suchkov*
INTRODUCTION 1
Thermodynamics 2
Laws of Thermodynamics 3
The Evolutionary Cell Biology in Terms of Energy Conservation 5
The Cell, the Forces of Gravity and other Physical Forces 6
The Cell, an Apparently Chaotic System 6
The Cell from the Point of View of the Entropy 7
Melanin, the Fundamental Molecule of Life and Evolution 8
Photo-protective Role of Melanin 8
Melanin and Other Pigments in Biology 10
Melanin Functions Described in the Literature 13
Melanin as an Energy Source 15
The Chemistry of Melanin 19
The Physical Properties of Melanin 19
Optical Properties 20
Indirect Evidence of the Unsuspected Property of Melanin 20
Kleiber Law ($E = M^{3/4}$) 22
Why Big Animals Use Energy More Efficiently Than Little Ones? 23
Melanin Functions in the Eye 24
How the Results Have Been Achieved 25
Photosynthesis 34
Photosynthesis in Plants 35
Photosynthesis in Humans 35
ATP is Not the Perfect Currency for Energy Cell 38
Energy Cost of Protein Synthesis 38
Human Photosynthesis and Its Potential Implementation in PPPM 39
Some Examples of Potential Applications of the Human Photosynthesis Enhancement 39
Age-related Macular Degeneration (AMD): Alzheimer´s Disease in the Eye ? 39
Neurotransmitters Role 42
CONCLUSION 42
Challenges of Current Healthcare Systems 43
Outlook 45
HUMAN AND ANIMAL RIGHTS 46

CONFLICT OF INTEREST 46
ACKNOWLEDGEMENTS 46
REFERENCES 46

CHAPTER 2 THE CELL BIOENERGETICS PATHWAYS BEYOND GLUCOSE AND MITOCHONDRIA AND THE INTRINSIC CHEMISTRY OF MELANIN 51
Arturo Solís Herrera, María del Carmen Arias Esparza, Ruth I. Solís Arias, Paola E. Solís Arias and *Martha P. Solís Arias*
INTRODUCTION 51
Background 52
Glycome 53
The Central Paradigm and Melanin 53
Clinical Contradictory Data 55
Melanin as Unsuspected Source of Chemical Energy 60
Energy Definition 61
Glucose: The Universal Building Block 64
The Regularity in Living Things 65
Glucose and Energy 66
Kidney Cortex and Chemical Energy 67
Adenosine, the Backbone of ATP 67
CONCLUSION 70
HUMAN AND ANIMAL RIGHTS 71
CONFLICT OF INTEREST 71
ACKNOWLEDGEMENT 71
REFERENCES 71

CHAPTER 3 WATER OF THE CEPHALOUS SPINAL FLUID- THE MAIN SOURCE OF ENERGY OF THE CENTRAL NERVOUS SYSTEM 74
Arturo Solís-Herrera, Graciela Landín-Miranda, Ruth I. Solís-Arias, Paola E. Solís-Arias, Martha P. Solís-Arias and *María del Carmen Arias Esparza*
BACKGROUND 74
Glucose: A Source of Energy? 74
Mitochondria 75
Cephalous Spinal Fluid (CSF) and the Physiology of the Choroid Plexuses (CP) 77
Human Photosynthesis or the Unexpected Capacity of Melanin to Split the Water Molecule 78
The Energy Supply to the Nucleus of Eukaryotic Cell Without Mitochondria or ATP 79
Melanin and the Central Nervous System 83
CONCLUSION 87
HUMAN AND ANIMAL RIGHTS 87
CONFLICT OF INTEREST 87
ACKNOWLEDGEMENTS 87
REFERENCES 87

CHAPTER 4 THE UNSUSPECTED INTRINSIC PROPERTY OF RESPIRATORY PIGMENTS TO DISSOCIATE, IRREVERSIBLE, THE WATER MOLECULE 89
Arturo Solís Herrera
INTRODUCTION 89
Oxygen Transportation 90
The Exchange Membranes 95
Hemoglobin 99
Hemocyanin 100

Chlorocruorins 101
Hemerythrin 101
Oxygen and Respiratory Pigments 101
Venous Oxygen Tension 104
Oxygen Tension and Exercise 105
Hemocyanin and the Transport of Oxygen in Marine Animals 107
CONCLUSION 108
HUMAN AND ANIMAL RIGHTS 109
CONFLICT OF INTEREST 109
ACKNOWLEDGEMENT 109
REFERENCES 109

CHAPTER 5 THE SOURCE OF ENERGY OF THE NUCLEUS IN EUKARYOTIC CELL IS MOLECULAR HYDROGEN AND HIGH ENERGY ELECTRONS 110
Arturo Solís Herrera, Arias Esparza, María del Carmen, Solís Arias Ruth I., Solís Arias Paola E. and *Solís Arias Martha P.*
INTRODUCTION 110
Light/Melanin/Water, the Human Photo-system 114
CONCLUSION 118
HUMAN AND ANIMAL RIGHTS 120
CONFLICT OF INTEREST 121
ACKNOWLEDGEMENT 121
REFERENCES 121

CHAPTER 6 MENTAL ILLNESS, ALTERATIONS IN VENTRICLE SIZE, AND HUMAN PHOTOSYNTHESIS® 122
Arturo Solís Herrera, María del Carmen Arias Esparza, Ruth Isabel Solís Arias, Paola Eugenia Solís Arias and *Martha Patricia Solís Arias*
BACKGROUND 123
CONCLUSION 131
HUMAN AND ANIMAL RIGHTS 132
CONFLICT OF INTEREST 133
ACKNOWLEDGEMENT 133
REFERENCES 133

CHAPTER 7 AGE-RELATED MACULAR DEGENERATION, THE ALZHEIMER´S DISEASE OF THE EYE AND THE INTRINSIC PROPERTY OF MELANIN TO 135
Arturo Solís Herrera, Ruth I. Solís-Arias, Paola E. Solís-Arias and *Martha P. Solís-Arias*
BACKGROUND 136
The Photosynthesis in Humans or Melanin´s Molecule Intrinsic Property to Splits and Reform the Water Molecule 138
Material and Methods 142
Therapeutic Results in Age-related Macular Degeneration 145
Monoclonal Antibodies Anti-VEGF in the Treatment of ARMD 153
Examples of Clinical Cases 158
Vitamins, Supplements and Anti-oxidants in the Treatment of ARMD 162
CONCLUSION 162
HUMAN AND ANIMAL RIGHTS 163
CONFLICT OF INTEREST 163
ACKNOWLEDGEMENTS 163
REFERENCES 163

CHAPTER 8 **ENERGY PRODUCTION, THE KEY ROLE OF THE MELANIN MOLECULE IN THE HUMAN BODY: IMPLICATIONS IN THE CONTEXT OF AGING** ... 167
Arturo Solís Herrera, María del Carmen Arias Esparza, Ruth Isabel Solís Arias, Paola Eugenia Solís Arias and *Martha Patricia Solís Arias*
THE ORIGIN OF LIFE ... 168
Cell Functions ... 170
ATP, is Supposedly the Energy Currency of the Cell ... 171
Mitochondria ... 172
The Not-possible Enzyme ... 174
Uterine and Umbilical O_2 Uptakes ... 175
Water and Glucose ... 175
The Source of O_2 for the Cell ... 176
The Cell Can Produce O_2 at its Own, *In Situ* ... 176
Oxygen is Toxic at Any Level ... 177
The Uniqueness of Melanin ... 179
CONCLUSION ... 184
HUMAN AND ANIMAL RIGHTS ... 185
CONFLICT OF INTEREST ... 185
ACKNOWLEDGEMENT ... 185
REFERENCES ... 185

CHAPTER 9 **THE UNEXPECTED ROLE OF MELANIN AS BIO-ENERGETIC MOLECULE AND PROF. SZENT-GYORGYI´S ENERGY CONCEPTS: HYDROGEN AS MAIN SOURCE OF ENERGY OF MUSCLE CELL** ... 187
Sergey Suchkov, Arturo Solís-Herrera and *Ruth Isabel Solís-Arias*
INTRODUCTION ... 188
Hydrogen as the Real Source of Energy of the Muscle ... 191
The Structure and Chemistry of Muscle ... 193
Muscle Changes After Death, Rigor Mortis ... 195
Molecular Hydrogen and Rigor Mortis ... 196
Differences Between Relaxation and Contraction ... 200
Bioenergetics and Heart Muscle ... 209
How a Bond-energy Could Lock Up in a Chemical Link Produce Work Somewhere Else? 210
High Energy Electrons ... 212
The Visionary Concept of Common Energy Levels of Prof. Szent-Gyorgyi ... 213
Lactic Fermentation ... 214
ATP and Energy ... 215
Melanin, the Unsuspected Bioenergetic Molecule ... 218
Optic Nerve in Young ... 219
Melanin Nearby the Optic Nerve ... 220
Melanin and Macular Diseases ... 221
Choroidal Nevi ... 221
Pigmentary Retinitis ... 222
Electromagnetic Radiation (Laser) Activates the Synthesis of Melanin ... 222
Melanin and Myopia ... 223
Melanin Releases Energy Symmetrically in all Directions ... 225
CONCLUSIONS ... 225
HUMAN AND ANIMAL RIGHTS ... 227
CONFLICT OF INTEREST ... 228
ACKNOWLEDGEMENTS ... 228
REFERENCES ... 228
SUBJECT INDEX ... 229

PREFACE

This book is the result of more than two decades of research, which began with a simple question: what can we do to try to stop the inexorable advance of the three leading causes of blindness in Mexico and that are also all over the world?

The answer was not simple, because despite my best efforts as an ophthalmologist, my patients were losing vision both for glaucoma, diabetes and macular degeneration by age. And the statistical information at the local level, regional and global showed that the incidence and prevalence of these diseases remained unchanged, which meant that the treatments available, by expensive, heroic, sophisticated or pompous to be called were not result.

Research we propose, initiated in 1990, was only intended to find morphological changes in vessels that emerge and penetrate (arteries and veins, respectively) into the optic nerve head, which can be used as indicators of early disease, enabling early treatment.

Nobody thought that the research would be more fruitful than we could even imagine. Twelve years later, in February 2002, was stunned at the results: the melanin was able to dissociate the water molecule.

At that time I was aware of that the result was important, even I thought that already knew many researchers; so it continued working on developing strategies that would allow me to modulate the activity of melanin, as years before understanding as acted, I had noticed that when melanin appeared to be active, the ocular tissues were in a better State than when melanin seemed "off".

It wasn't until 2005, listening to a speech by the then president Bush, who literally said "We need substances that remove hydrogen from water to enter fully into the era of hydrogen".

And I kept thinking about: because not used melanin? Greener want it? We have it all. I began to read to trying to find the answer, and a few months later I came to the conclusion: do not use it because they have not realized. So I decided to start the procedure of patent.

Arturo Solís Herrera
Human Photosynthesis® Study Center
Sierra del Laurel 212, Colonia Bosques del Prado Norte
Aguascalientes, 20127
México

DEDICATION

Dedicate this book to all those who in one way or another, allowed to be perform. And especially to my family all

INTRODUCTION

PRESENTATION

Few molecules have been so reviled than and despised as melanin. Some think that melanin is a stigma that tends to disappear as the civilizations advance. It seems that a greater amount of melanin, greater contempt. The least that is said, is that melanin is a simple solar filter, which only protects us from dangerous UV rays, hypothesis proposed in London in 1820 by Sir Everard Holmes. To date, the idea is still present, even among dermatologists.

Melanin are high molecular weight pigments, ubiquitous in nature, with a variety of biological functions. In mammals, melanin is found in skin, hair, eyes; brain, eye, and in many internal tissues [1]. The wide variations in hair coloration in mammals are all the result of the melanin paint brush.

Animals with dark fur usually have few pigmentations in the skin. Oppositely, polar bears have black skin under their white fur. Melanin´s thermoregulatory action supposedly is due to tyrosinase, the main enzyme involved in melanin synthesis; is temperature sensitive compound and active only in cooler regions of the skin.

In amphibians, the dispersion and aggregation of melanophores, make the animal dark or pale, respectively. These particular responses are under hormonal control. Theoretically, in plants, melanin and related pigments may be metabolic by-products. The lignin of the wood is very similar to melanins.

Careful assessment of minimal erythema dose (MED) has been shown that the protective role of skin pigmentation is not merely of a sun screen due to a sunscreen would raise the threshold. Paradoxically, melanin´s sunscreen efficiency is greatest at biologically irrelevant wave lengths. The sun protection factor (SPF) of melanin is 2, the same as a solution of copper sulfate at 2% concentration.

Many chemical and drugs are bonding and retained in tissues that contains melanin. This binding property can be both detrimental and beneficial [2]. Drug-melanin interactions may be a two-edged sword. Melanin has defied optical characterization, and its absorption pattern looks more like inorganic material than organic [3]. This means that electrical properties of melanin can be described as exotic.

Melanin is known to have a large capacity to absorb atmospheric moisture and other polar gases, but the mechanisms involved are poorly understood. So far, there is no satisfactory knowledge of the molecular structure of melanin pigment granules. Melanin is a bio-macromolecule, and one of its most unusual properties is its persistent electron spin resonance (ESR) signal, which means free radical centers presence in the material [4]. There is a very little quantitative data related to redox properties of melanin, probably because the technical obstacles to make experiments with the material.

Heating of melanin or irradiation with light, induces transient melanin radicals [5].

An intriguing finding of several studies on human hair melanosomes [6] and cow ocular melanosomes [7] is that while the amounts of diverse heavy metals ions bound to red-hair and black hair melanosomes are significant and rather similar, the Fe (III) content is four times higher in the red hair melanosomes. The explanation, at this time, is not clear.

The photo-dynamics of melanin is very fast, in the range of Nano and Pico seconds, and is responsible for almost complete conversion of the energy of the absorbed photons into heat, theoretically. The photoreactions of melanins are so far, considered responsible for the generation of superoxide anion and hydrogen peroxide, and supposedly, to the photodegradation of pigment [8]. The melanin mediated photo-reaction, can reduce molecular oxygen; although the exact origin of involved electrons has not been unambiguously determined [9].

Prolonged aerobic irradiation of melanin with UV or even short-wavelength visible light, leads to melanin bleaching [10]. The observed photobleaching explained theoretically by oxidative photo-degradation *via* ring-opening reactions [11]. Melanin is considered an intrinsically insoluble pigment, but when is photo-degraded, appears a distinct solubilization [12].

It is unclear whether *in-vitro* studies of melanin have any biological implication.

The above data are very interesting, but to date it does not seem possible to correlate between them to allow the proper structuring of the knowledge, which, at this moment, seem loose.

The widespread presence of melanin in nature reflects its importance in biology, but the substantive technical obstacles to study it in the laboratory, have constituted a serious difficult to advance in the complex problem of the biological role of melanin.

So, the big question remains: which are the main function of melanin, in example: in the eye?

Melanin Functions in the Eye

It is generally accepted that melanin in the skin and eye (Fig. **1**) acts as natural sunscreen, absorbing and scattering solar radiation, mainly the UV rays and short wavelength visible photons. A correlation between the UV-induced erythema and sunburn of human skin, and its constitutive pigmentation is observed [13]. Skin susceptibility to photoaging and skin cancer (solar radiation-related) inversely correlates with the amount of melanin in the skin [14].

Through unknown mechanisms, melanin was shown to protect against induction of DNA lesions by UVB, in cultured melanocytes [15]. The main cause of blindness in cold countries, age-related macular degeneration, is more often found in individuals with lower content of uveal and skin melanin [16]. The mechanism of photoprotection by melanin at molecular and cellular level is poorly understood. The very efficient non-radiative de-excitation of melanin or energy dissipation, there has not unraveled [17].

The molecular nature and behavior of melanin under normal conditions and age-related changes is unknown. It is necessary to solve the structure-property problem.

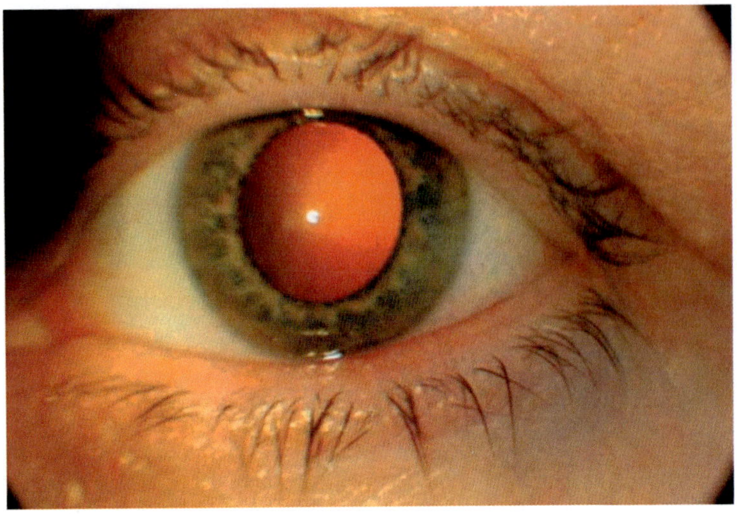

Fig. (1). Melanin is present in the eyelashes, in the skin, in the eyebrows, in the iris, *etc*. Melanin is especially noticeable on the edge of the pupil.

Melanin is present in the human eye wherever we look for it (Fig. **2**). Melanin is not unique to the superficial tissues of the human body or animals.

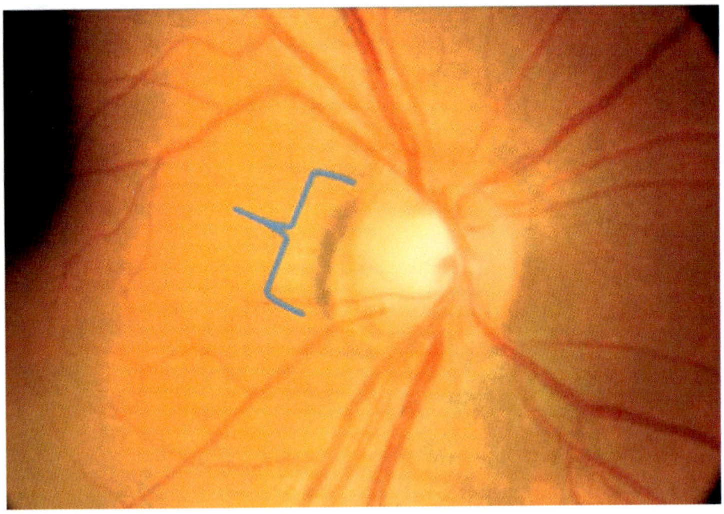

Fig. (2). The picture shows the optic nerve of a clinically healthy person. However, melanin is present at the edge of the nerve (blue line).

It has long been said that melanin is a ubiquity molecule, as it is one of the pigments found in all living beings. Its apparent stability, although it is considered by some researchers as a disordered system within the molecule; it was one of the main obstacles to be able to give it a role in biology.

So far it has been considered primarily as a simple solar filter, but its location inside the body, as in the midbrain, the inner ear, and the interior of the eye (Fig. **3**), cast doubt about its photo protective role.

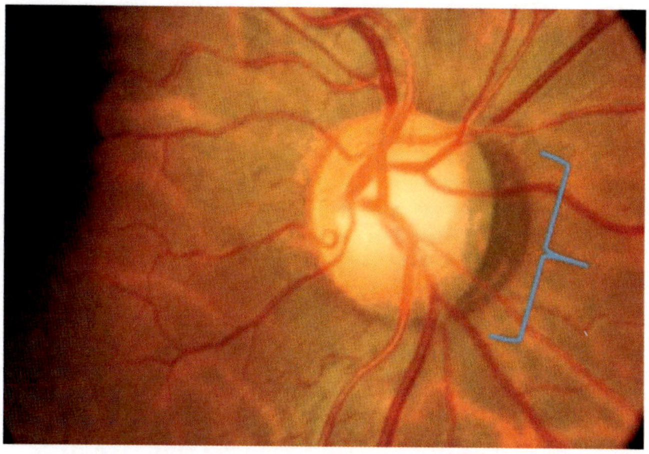

Fig. (3). The melanin covers the entire interior of the eye, it upholsters it completely, but it is perceptible around the optic nerve, as in the photograph, where the melanin is pointed out by a blue stroke.

The uveal tract completely covers the inside of the eye of the human eye (Figs. **4**, **5**), and is a pattern that is repeated in all species. The only function currently attributed to the high melanin content of the choroid layer is to absorb excess light, so that the light that penetrates the eye is not reflected in the interior, which contributes to a better image quality.

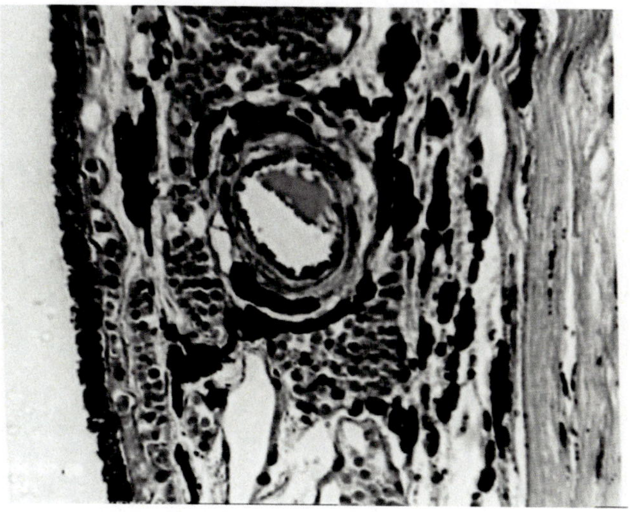

Fig. (4). The microphotography corresponds to the uveal tract of the eye, of the Greek: grape color. It is the most pigmented part of the eye as it contains 40% more than the skin. The uveal tract or choroid layer completely covers the inside of the eye.

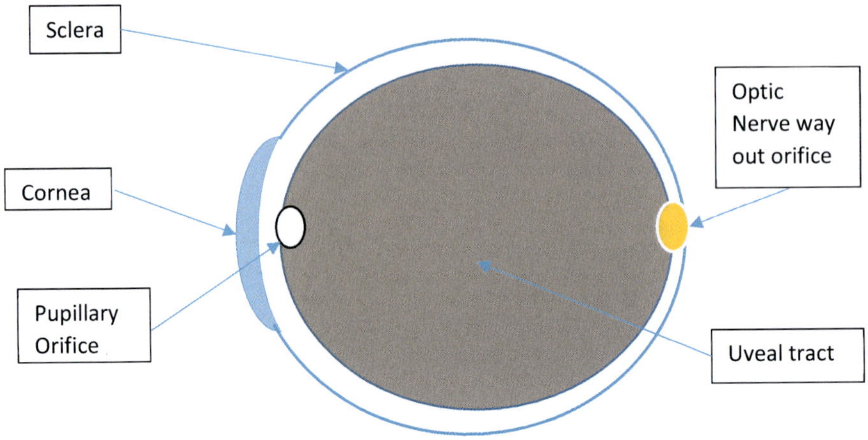

Fig. (5). The uveal tract fully upholsters the inside of the eye, it has only two main holes, one of which corresponds to the pupillary orifice and the second corresponds to the exit hole of the optic nerve. The uveal tract resembles a hollow sphere on the inside.

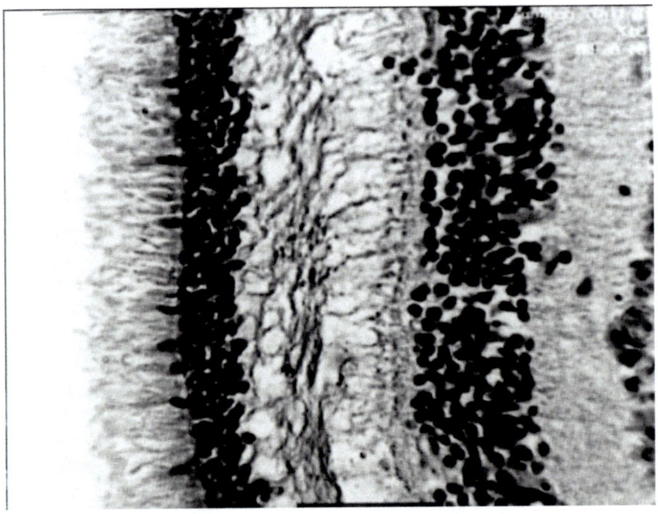

Fig. (6). The human retina and in mammals in general, is a very thin layer, almost transparent tissue, which is attached to the uveal tract.

The adhesion of the retina to the uveal tract is not understood to date, as it is a dynamic process, rather than anatomical.

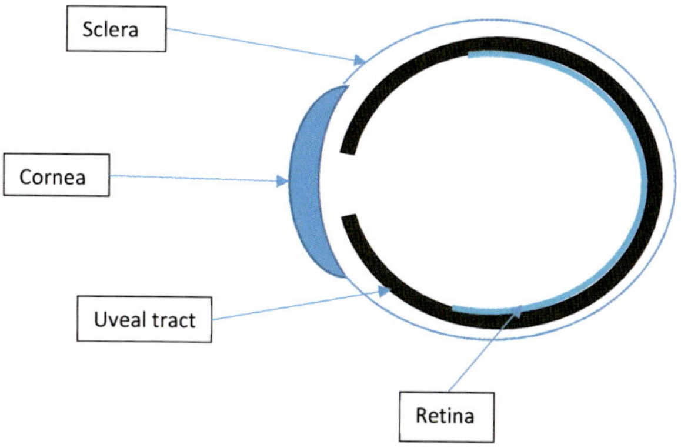

Fig. (7). The sclera of the eye is a fibrous layer that forms an envelopment around the choroidal layer.

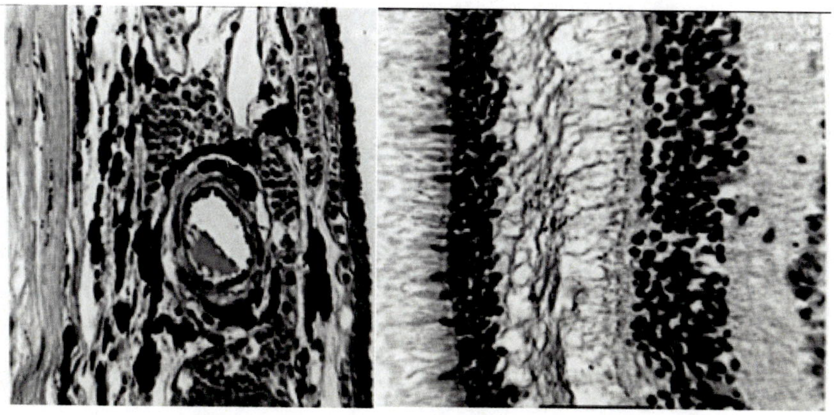

Fig. (8). Photographic composition showing the position of the tissues described above: retina (right) and uveal tract or choroidal layer on the left.

The photoreceptors (cones and rods) are slightly immersed in the first layer of the uveal tract (Figs. **6-8**), which is called the retinal pigmented epithelium, although anatomically it belongs to the choroid layer.

The current concept of melanin functions in the eye is limited to absorbing excess light, allowing a clearer and more precise vision. The foregoing has been a sufficient explanation for decades. But an anatomical pattern that is repeated in all humans and in the species, that have eyes, raises doubts about such a simple function of melanin. It is possible that the biological role of melanin is more extensive, more fundamental than it is currently attributed.

It is precisely the purpose of this book to delve into it.

The Omnipresence of Melanin in All Biological Realms (Figs. 9-93)

It is a very rooted idea that the main function of melanin is to protect us from ultraviolet radiation. Its main function, from quite simplistic point of view, is to protect us from excess light.

But in nature, our prejudices have you carelessly. Surprisingly the presence so extensive of melanin in nature, and more surprising that has gone unnoticed for so long.

Fig. (9). The squirrels, the presence of melanin is undeniable, it is enough to observe the eyes, fur and tail.

Fig. (10). Fruits and seeds are no exception, as they also contain melanin.

Theoretically, UV radiation induces skin cancer, but in plants and seeds, cancer is nonexistent.

Fig. (11). Melanin is present in all biological realms. In the skin of the rhinoceros, the presence of melanin is substantive and in its full extent.

Fig. (12). The omnipresence of melanin is observable in any species. In the photograph we can see it in the skin of the pigs, and in the lignin of the trees.

Fig. (13). The skin of the American bison shows a great concentration of melanin.

Fig. (14). The buffalo skin, in its full extent, also contains significant concentrations of melanin.

Fig. (15). The dromedary, like many species, presents remarkable concentrations of melanin in its entirety.

Fig. (16). In turtles, melanin is found not only in the skin, but also in its shell.

Fig. (17). Nature is concerned that melanin is present in all living things. In the case of the turtle it is not only in the skin, the shell, the eyes, but also in the fingernails. Nature only insists on important things.

Fig. (18). In the Gila monster, the skin has stripes where the amount of melanin is greater than in others. and the presence of melanin is re-fulfilled.

Fig. (19). The distribution of melanin in the skin of the different vertebrate species is characteristic. But the presence of the master molecule is always constant.

Fig. (20). In reptiles, the presence of melanin is no exception. Its presence is universal, so its biological role must be beyond a simple solar filter.

Fig. (21). In amphibians, the presence of melanin appears to be discrete, but even the brown hue is due to melanin.

Fig. (22). The distribution of melanin can change according to the species of reptile, but invariably it is always present.

Fig. (23). Melanin, distributed in a different way, but it is always part of the living being in question.

Fig. (24). In the different species of reptiles, melanin is distributed in a characteristic form for each species. But it will always be present in one way or another.

Fig. (25). In the carapace and skin of the turtles, melanin is a constant, despite the different distribution.

Fig. (26). The appearance, and electron properties of melanin, change significantly when they are hydrated. To date, the explanation was not known.

Fig. (27). The presence of melanin in the tortoise shell is important. Turtles that are replaced by a plastic shell, for example, after an accident, do not survive for long.

Fig. (28). The appearance of melanin changes when hydrated, which can be seen at first sight.

Fig. (29). The amount and distribution of melanin in reptiles is different for each species, not their constant presence.

Fig. (30). In chameleons, also melanin is always present.

Fig. (31). In reptiles (and in the trunk of trees) can vary the size, length, thickness, weight, *etc.*, but the omnipresence of melanin is invariably in both plants and animals.

Fig. (32). In the iguana, melanin also acts of presence, and in some areas of the skin, the concentration of the pigment is remarkably high. The insistence of nature to place melanin in all living things definitely has another explanation, beyond the merely photo protection.

Fig. (33). Water and melanin are constant elements in any way of life. It is said that water is life, but we could add that light, melanin, and water are essential elements for the miracle of life to happen.

Fig. (34). The need for light in living things, which is observed in daily life, does not have a satisfactory explanation to date. How to concatenate melanin and water in that intuitive need for light of living beings is one of the purposes of this book.

Fig. (35). The need for light is intuitive. Both humans and animals are looking for light.

Fig. (36). What is the reason that the tortoise shell contains significant amounts of melanin?

Fig. (37). Melanin is widely distributed in nature. The photograph shows his presence in the reptile and in the trunk of the tree.

Fig. (38). Melanin present in both the reptile and the tree trunk.

Fig. (39). The designs on the skin of the reptiles, are fascinating, and in all of them the melanin is the basis

Fig. (40). Light, melanin, and water. In order of abundance in the universe.

Fig. (41). The beauty of melanin-based designs that creates nature in the skin of reptiles is immense.

Fig. (42). The magnifying of the previous photograph, highlights a pattern that is repeated in the animal Kingdom: the organs of the senses have a higher concentration of melanin than the rest of the tissues observable to the naked eye.

Fig. (43). The innate search for light is something that was not fully understood to date. It happens in all animal and plant species.

Fig. (44). Melanin is the same in all living things. The differences are reduced to the size of the granule, the orientation of it, the tissues that surround it, the number of granules; *etc.*

Fig. (45). The macro and microscopic appearance of melanin, as well as its electronic properties, change dramatically depending on the amount of water. It was a phenomenon that had not been able to be explained to date.

Fig. (46). If the species is, in evolutionary terms; of recent or previous appearance, is not an obstacle to the presence of melanin.

Fig. (47). The pattern of distribution of melanin appears to recur regardless of whether the animals are homeotherms or poikilotherms.

Fig. (48). In all biological realms the presence of melanin is indisputable. The tarantula is a palpable specimen.

Fig. (49). In the case of amphibians, in this case the axolotl, melanin returns to act of presence. Its importance in biology must be enormous given the insistence of nature.

Fig. (50). The axiom is also fulfilled in the fish. Melanin seems to be an indispensable element for life.

Fig. (51). The skin of the iguana presents a coloration that is explained, again, by the presence of melanin.

Fig. (52). and the distribution pattern of melanin is repeated. Note that the concentration of melanin is higher in the organs of the senses, in this case the eye.

Fig. (53). Even the terrifying Komodo dragon, the presence of melanin seems to be indispensable.

Fig. (54). And melanin seems to have a special predilection to locate in the organs of the senses and in areas close to them.

Fig. (55). In the case of deer, the special location of melanin in the organs of the senses, note the nose and eyes.

Fig. (56). Evolutionarily speaking, food originated at the same time as living beings or perhaps before. It is more feasible to think that life originated in the midst of an abundance of food substances than the opposite. Carbohydrates, lipids, amino acids, vitamins and minerals that food contains, are used by living organisms to replenish biomolecules that are worn or discarded with normal metabolism.

Fig. (57). The melanin that is observed in the skin of the giraffe, is a type of melanin that is called Pheomelanin, which contains, characteristically, abundant disulfide bonds.

Fig. (58). The organs of the senses seem to require a greater amount of melanin to function properly. The reason for this was not so far understood.

Fig. (59). The melanin present in the chimpanzees is indistinguishable from the human. Apparently, a greater amount of melanin is an advantage. The chimp has the strength of five men.

Fig. (60). Melanin is present in any living being, whether an orangutan or a chimpanzee.

Fig. (61). In the Okapi (Okapia johnstoni), mammal of the family Giraffidae, melanin makes its accustomed appearance.

Fig. (62). The sable antelope is not exception. Melanin in the fur with higher concentration around senses.

Fig. (63). It is a constant behavior of melanin, the highest concentration around eyes, nose and mouth.

Fig. (64). Hippotraginae, orix. With its characteristic melanin distribution.

Fig. (65). Pongo pygmaeus. The reddish hue is due to melanin with high content of sulfur.

Fig. (66). Elephantidae. The largest terrestrial animal that exists today, and the melanin present.

Fig. (67). The photograph shows that the hair is scarce and the skin thick, but the presence of the melanin is evident.

Fig. (68). The presence of melanin is essential for life. And melanin has a great affinity for calcium.

Fig. (69). The amount of light influences our perception. Although it is the same melanin in both specimens, the color seems different.

Fig. (70). Melanin tends to approach energy sources.

Fig. (71). In general terms, the colors observed in the different kingdoms are explained by the presence of melanin or its combination with other elements, for example, some metals.

Fig. (72). The presence of melanin is constant in all living things. Food presents greater variations, given that, mainly, they are used as precursors of the biomolecules that make up us and that the body synthesizes.

Fig. (73). Peafowl: The iridescent coloration of the large train of Peacock, blue and green plumage; they are the result of combining melanin with metallic elements, but melanin remains the main basis of this pigmentation.

Fig. (74). In the side portion of the peacock, as well as in the distal portion of the plumage, melanin can be easily observed.

Fig. (75). The back part of the peacock, the presence of the dark pigment is more noticeable.

Fig. (76). In this photograph, it is seen that the plumage closest to the body of the peacock is dark, denoting the presence and unusual importance of melanin.

Fig. (77). The appearance of melanin is similar even among distinct species. From a chemical point of view, they are indistinguishable from each other.

Fig. (78). The amount of melanin present in the fur and skin of the gorilla is substantive. It covers the whole animal. Why so much insistence on nature?

Fig. (79). The silver tenderloin of male adult gorillas is because melanin, as it progresses in age, begins to have a less homogeneous distribution than young animals.

Fig. (80). The homogeneity of the distribution of melanin in the skin and fur of animals, is a common characteristic of young specimens.

Fig. (81). The melanin of the gorilla is indistinguishable from the melanin of the humans. The higher concentration of melanin seems to be an advantage for life, as the world champion Weightlifting, male; It has a record of 200 kg, while a gorilla, male, adult; can lift 4 tons.

Fig. (82). A good test of the wide presence of melanin in all living beings, we have in this photograph in which we can see the dark coloration of the rhinoceros and the plumage of the dove.

Fig. (83). It is important to draw attention to the constant presence of melanin in individuals of the same species. The invariably presence of a molecule so far considered as a simple solar filter, in all living things, reflects an unsuspected importance.

Fig. (84). It can be said that the melanin molecule is the same in all species, and its anatomical disposition is also very similar. It is to be expected that its function is also analogous: to transform the visible and invisible light into chemical energy by means of the dissociation of the water molecule, like the chlorophyll in the plants.

Fig. (85). The beauty of the coat with abundant melanin is something that we recognize instantly. There is not much difference with the beautiful hair of the girl in the next photo.

Fig. (86). Melanin is a molecule whose complexity far exceeds our capacity for abstraction, hence to date, it is not possible to distinguish, by chemical tests; between the melanin of the distinct species.

Fig. (87). The disposition of the melanin in the different species is characteristic of each one of them. The photograph shows the typical disposition of melanin in the zebra.

Fig. (88). In humans, the amount of melanin in the skin is lower than in other species. The photograph shows a person who could be classified as photo-type IV, dark hair and dark eyes.

Fig. (89). Melanin is also present in plants, but is notable mainly in the trunk.

Fig. (90). In the plumage of birds, melanin is also a constant characteristic.

Fig. (91). The color of the trunk of the trees is like the coloration of the skin of some species, and the reason is that, in both cases, is due to the presence of melanin.

Fig. (92). Melanin behaves in a surprisingly constant way in all animal or plant species, because when it is present, it tends to approach the energy sources and surround them.

Fig. (93). The melanin in the tree trunk is as surprisingly constant as its presence in the skin, fur or feathers of the animal species. Melanin can be found in any being alive, at all latitudes.

Fig. (94). Wherever that the look is directed, we will find the melanin.

The melanin molecule is very similar in all species, and also its function: transform light energy into chemical energy dissociating the water molecule (Figs. **94 - 96**), as chlorophyll in plants.

Fig. (95). The distribution of melanin in the skin of marine mammals, responds to the light source.

The observation that led me to discover the inherent capacity of melanin transform the visible and invisible light to chemical energy through the dissociation of the molecule of water, such as chlorophyll in plants, was precisely the omnipresence of the melanin in the vicinity of the optic nerve, in the almost six thousand patients that were reviewed during the observational study that began in 1990. I may not have been necessary to reach it, if it had been repaired in the also omnipresence of melanin in all biological kingdoms.

Melanin tends to approach the sources of energy, in this case, to the Sun.

Fig. (96). The beautiful picture of the steed, begins by the melanin skin contains in abundance.

Fig. (97). Bighorn sheep skin because its different shades melanin, whose granules can be of different sizes, different orientation, differences in the molecular weight of pigment, however, its function remains the same: production of chemical energy.

The biology of life is the same since the beginning of time, and even, in prehistoric specimen melanin always is found.

Fig. (98). In ravens, the amount of melanin is very marked.

Fig. (99). In leopards, melanin is evident on the skin and gums.

The obscure's and the Red of blood and skin support to Darwin, who said that although there are millions of species, life is just one. The constant presence of melanin suggests a dominant role not only in life itself, but from its origin.

Fig. (100). In the seal also melanin? Of course, and their function is the same: transforming the visible and invisible light into chemical energy.

Regardless of without are cats or dogs, melanin is presence by supplying chemical energy.

Under certain conditions, the aspect of melanin is very similar, even among reptiles, and plants.

Despite the huge differences between different species (Fig. **100**), the vital principle is the same: the chemical energy that emanates from the melanin.

Life is possible at the ends, thanks to the extraordinary stability of the melanin. Demonstrated: 170 million years.

The amount of melanin in the skin is determined mainly, by the amount of light. In places of plenty of sunshine, the amount of melanin increases for seize this light energy, but at the same time regulates the amount of light which penetrates inside the body, since up to bone and muscle need light. In dim light, such as Northern Europe, the amount of melanin in the skin is less so that it can pass and pass through the tissues, nourishing them.

Fig. (101). Did they think scholars, that despite observing such amount of melanin and in different biological kingdoms; they delegated their role to a mere sunscreen?

Feeding is just exchange of biomass or more precisely of carbon chains. Energy comes from the light. In some specimens, melanin is especially contrasting. And their function is the same wherever you are. Even in birds, in their chicks, melanin cannot miss. Nature just insists on important things. Even in specimens of unknown species that eventually found on the beaches, the presence of melanin occurs without exception.

Fig. (102). The melanin present in the rhino and readily observable in the skin, is indistinguishable from the melanin present in other species, including the human.

Fig. (103). Plants also have melanin, mainly on the trunk and roots.

Fig. (104). Melanin forms masses of the size of galaxies in outer space. And their behavior is characteristic, tends to approach energy sources and surround them.

The huge roots of trees also require melanin to convert light into chemical energy (Fig. **103**).

In seabirds, seems to be repeating the distribution of melanin in some marine species such as dolphin and shark, because melanin is especially abundant in the upper part, which is the most exposed to the rays of the Sun, but not with the purpose to protect from UV rays, but to make the most likely to be transformed into chemical energy light energy.

Fig. (105). Fungi are no exception, also based their biology in the melanin.

Melanin is especially abundant in the organs of the senses. At any age, the presence of melanin is indispensable. Melanin is required since the initial stages of gestation. It seems no one millimeter of skin or furrow in which melanin was not present. So important is its presence.

In some species, the amount of melanin present seems to be especially notable, although is carefully regulated by around 200 genes, which determine the amount and nature of the molecule, according to factors such as the environment. The presence of melanin in all species, appears to be out of the question.

Fig. (106). Polar bear fur is dark due to the abundant melanin, perceptible in the nose. The hairs seem to function as optical fibers that transmit light to dark skin, and at the same time form a kind of insulation for cold.

Fig. (107). In some birds, the presence of melanin jumps at the sight. It is difficult to believe that, in spite of the insistence of nature to melanin in all living things, its main function has been neglected for centuries.

Fig. (108). Even the tiny moth uses the same principle: melanin as source of energy.

Reddish melanin is low molecular weight, the dark, is of high molecular weight (million Daltons).

Nature does not make distinctions between species. Melanin in each and every one, without exception (Figs. **104-120**), and its main function is invariable: transform light into chemical energy by dissociation and re-form of the water molecule.

Fig. (109). While the very obscure-haired dogs are relatively frequent in our daily lives, the importance of the melanin had gone unnoticed.

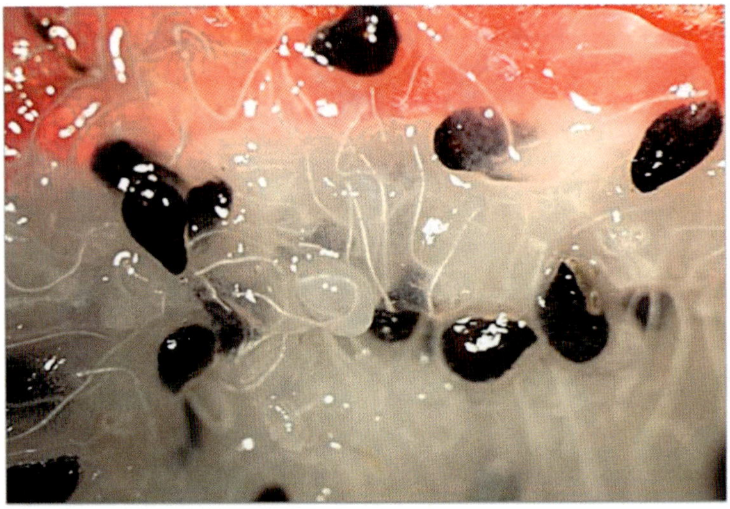

Fig. (110). Melanin is in all living beings, plants, animals, and others. In the case of Pitaya seeds, the presence of melanin in the seeds is beyond doubt.

Fish is another clear example of the omnipresence of melanin, and its remarkable dominance at the top, is consistent with the fact that the light always comes up.

Despite the remarkable presence of melanin, both human, tree trunks; as well as the tortoise, *etc*., its function had been relegated to the of a mere solar filter.

Turtles, like any living being, they obtained the chemical energy that require, separating the water molecule, by melanin, which is especially visible in the shell and skin. The need for breathing is mainly to expel CO_2.

Fig. (111). Stunning Bengal tiger also requires melanin to live. Dark Striate show, although melanin is really in all the skin, and the different shades of melanin, are due to variations in the molecular weight, the orientation of the molecule, the nature of the tissues surrounding it, combined with some metals; *etc*.

Fig. (112). The process of transformation of light energy into chemical energy is the same in plants and animals, but melanin in the plants is in the trunk, the books call it lignin. The lignin, a kind of melanin; withstands extreme conditions than chlorophyll not.

The shell of the tortoise possesses a remarkable amount of melanin, which is the same: transforming light into chemical energy through the dissociation of the water molecule. The behavior of turtles is not casual, they are exposing the melanin in the shell for energy.

Fig. (113). The milky way covered in significant part by the so called cosmic dust, that is indistinguishable from neuro-melanin. Cosmic Dust covering a significant part of the milky way.

Fig. (114). The behavior of the water droplets is influenced, in a substantive part; by the melanin of the plant.

Fig. (115). Melanin has variations in color depending of the amount of sulfur, size of the granule, orientation of the molecule, associated structures, *etc*. In the picture, the cat has a variant named pheomelanin, with content of sulfur.

Fig. (116). The disposition of the feathers of the birds, allows that during the flight they capture the molecules of water of the atmosphere, and thanks to the flapping, these molecules of water are directed towards the central part of the feather which is hollow. Gradually the water droplets fill the central hollow part, and the movement of the wings makes them go towards the skin of the bird, which normally has a high content of melanin.

Fig. (117). Once melanin and water come into contact, it begins the transformation of light into chemical energy by dissociating the water molecule into the melanin. The above explains why birds can fly thousands of miles without stopping, without drinking water or food; and when they reach their destination, they've almost lost weight. The classic example is the Arctic swallow. The intraocular organ named pecten is heavily melanized.

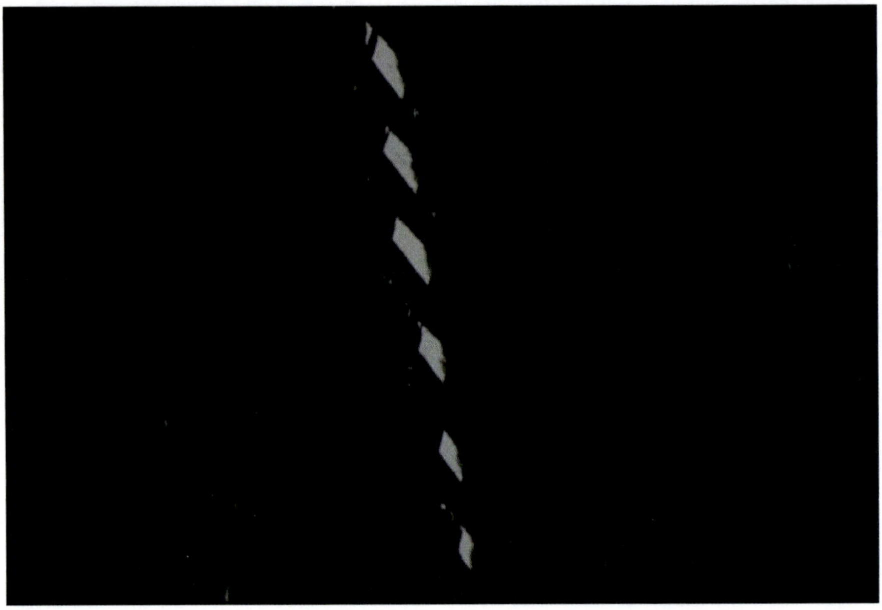

Fig. (118). Melanin can absorb all wavelengths of the electro-magnetic spectrum, from radio waves to gamma rays. The mechanisms of this astonishing fact is not understood. Therefore, it is said that melanin is the darkest substance known.

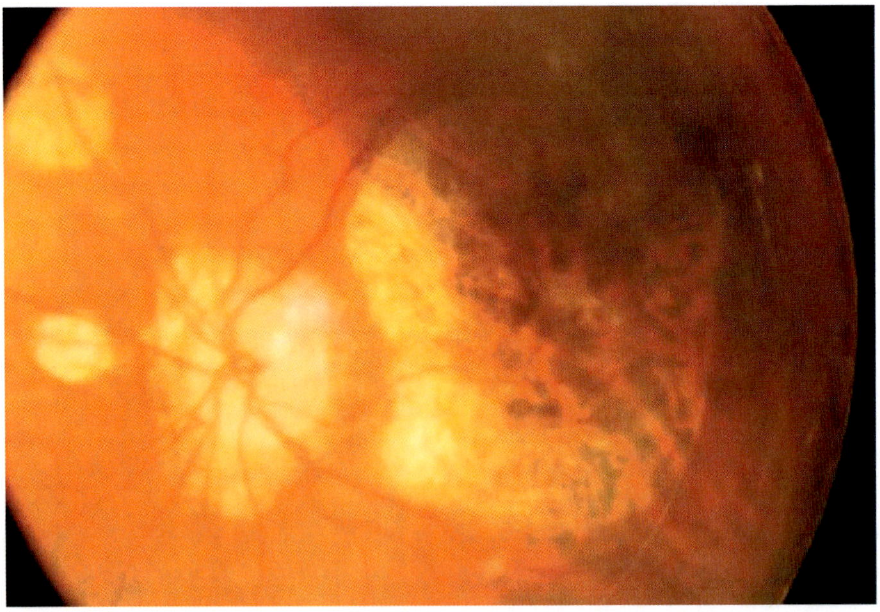

Fig. (119). The picture shows the retina of a living patient. The constant presence of melanin in the retina and around the optic nerve in all patients, detected in an observational, descriptive study; About the three main causes of blindness, initiated in 1990 and completed in the year 2002; It allowed us to detect the unexpected intrinsic ability of melanin to dissociate the water molecule, such as chlorophyll in plants.

Fig. (120). Melanin is present in plants, animals and insects. In the photograph a cricket (Gryllidae) is seen and at first sight there are areas where melanin is in greater quantity.

Fig. (121). Melanin tends to approach the energy sources and surround them. The pattern that we observe in the cat of photography, is repeated in other species, such as whales, dolphins, sharks, *etc*.

The omnipresence of melanin in all living things, in nature in general, and even in stellar space has gone unnoticed. Its function as a simple solar filter gained much acceptance since it was proposed in the eighteenth century, which difficult its study. On the other hand, melanin is very difficult to study in the laboratory, as it is said to resist the available study techniques. Incidentally, we study it in one of its natural locations, in the human eye (Figs. **121-123**); And that made the difference.

Melanin, the First Self-replicating Molecule

The alternative chemistries to the origin of life, must fit functional requirements from the standpoint of first principles. The ignorance of the totality of reactions involved in the creation of successive generations of increasingly complex prebiotic organic molecules is substantive.

The nature simple-to-complex of chemical self-organization, suggests an evolution from broadly ranging consecutive chemical trial and errors, were some pathways will work. However, this kind of chemical evolution requires the existence of a global chemical engineering system, which at the system level, amplify, at least theoretically; the probabilities of effective stochastic self-organization at the molecular level.

But the theoretical requirements of such a system is too long, for example: the likelihood that appropriate reactions happen have to be significant, the necessary processes have to be highly efficient and fast; the system to be able to explore continuously different routes possible chemistry as well as the mechanisms involved; that the system to be able to select only what is good, evil discard it; and be sealed to some extent, so that it is capable of retaining materials within the system total.

It would seem a race between a complex organizational structure that it self-building and at the same time self-organizing *versus* its tendency to dissipation due to entropy. The functional requirements of this ideal system, would be something like: ability to select and concentrate necessary appropriate compounds as reactants, stabilization and coordination of these reactants, able to control power and synthesis (directed synthesis); some stage products become the reactants of the next stage, and all of the above should occur in an appropriate geochemical and geophysical environment.

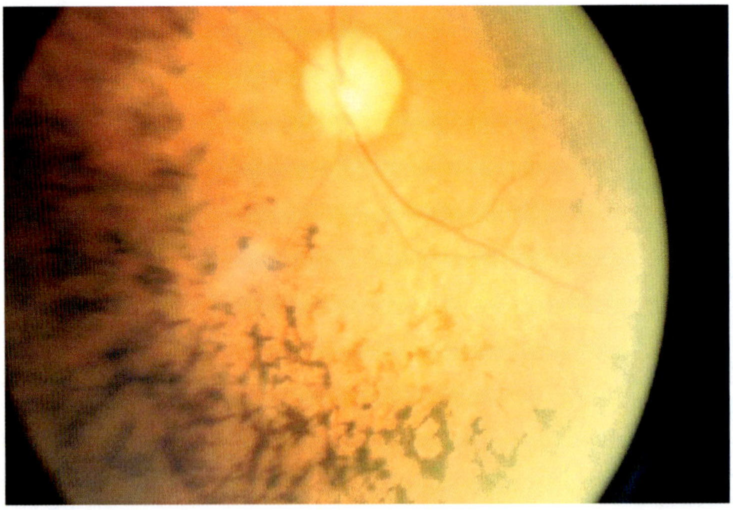

Fig. (122). Melanin is altered in most diseases of the retina, before it was thought to be secondary to the initial problems in the retina, but now we understand that first alters the generation and distribution of energy from melanin and then they alter the tissues that depend entirely on that energy.

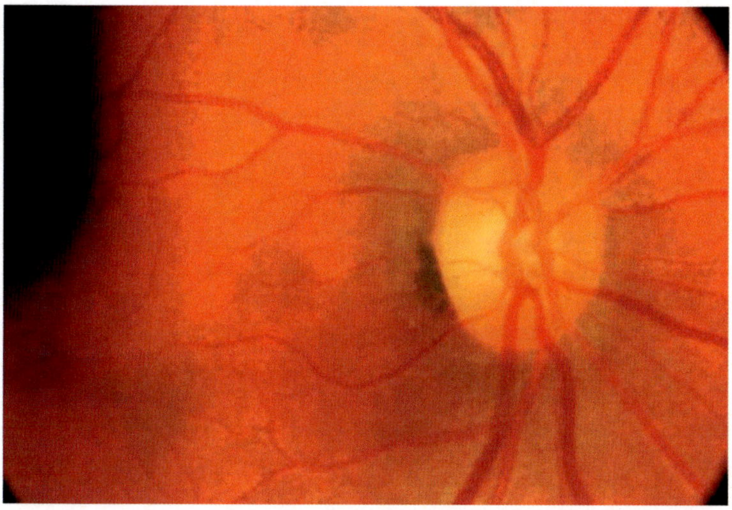

Fig. (123). The constant presence of melanin near the optic nerve, in the 6000 patients that were included in the study, over twelve years (1990-2002), was the first data that led us to discover the unsuspected intrinsic property of melanin dissociate the molecule of water in its gaseous components, hydrogen and oxygen, such as plant chlorophyll.

But reality has shown that difficulties in implementing a system like this, are formidable, as the chemistry of carbon, water-based, requires microenvironments that promote both heterogeneous reactions and dehydration. This, because the polymerization of biomolecules, required, almost universally make a H^+ and an OH^- scrapped at each end of the links in training, which means that, thermodynamically speaking, the expulsion of these elements isolated or combined (water), in an aqueous environment is very difficult.

This means that the pure environmental aqueous, is one obstacle rather than an advantage, on the way to the complexity of organic molecules. The possibilities are various, including the emergence of microenvironments that had added each other and in the end the complexity, against all odds entropic; it predominated. Or rather than processes that are part of a broader organization, larger; It led to the conditions at a microscopic level.

Theories on the matter are entertaining: *i.e.* the cycle of the bubble-aerosol - rain, also called the bubble hypothesis, model bubble Lerman, the hypothesis of the bubble in solution; where is examined thoroughly the physicochemical properties of interfaces, as well as cycles that could, but in the end, did not get anything conclusive, in spite to be called as: a universal planetary hydrology cycle.

About the appearance of prebiotic compounds remains outstanding questions.

It is for that reason that the theories about the origin of life are so far; just theories, because they still do not solve the basic problem. Another ingenious theory, is that the organic material that gave rise to life on earth could come by a series of collisions of comets. But trying to concatenate the few findings and numerous theories, the result is similar: uncertainty.

Melanin, the Molecule Base of Biological and Planetary Systems

Apparently, the intrinsic properties of melanin, it comes to solve so far intricate questions about the origin of life.

- Ability to select and concentrate necessary appropriate compounds as reactants, how is it possible?
- Melanin delivering power in a very consistent way, constant, incessant, as it takes place both day and night. The ranges that the H_2 and thc e^- derived from it are also surprisingly accurate; so, the effects that happen in the close environment is also similar, since different atoms or elements that are influenced by this chemical energy (H_2 and e^-) behave in a similar way all the time, this is forming the same compounds, with the same of primary, secondary, tertiary and quaternary chemical characteristics.
- Stabilization and coordination of these reactants. How is it possible?
- In the presence of visible and invisible light and water, the melanin provides chemical energy for thousands or millions of years. Which speaks of an enormous influence in terms of consistency of products it generates, as products are repeated and again throughout eons of years.
- Able to control power and synthesis (directed synthesis). How is it possible?
- The energy that emanates from the melanin, in the presence of light and water, is generated with accurate characteristics repeatedly, both day as at night, for millions and millions of years, carbon chains that are formed because of this as consistent energy, are also consistent. What effect has the chemical energy of the melanin on an isolated carbon atom is somewhat different to that happens when present 2, 3, 4, are or more carbon atoms, but the result will always be the same for millions and millions of years.
- Stage products become the reactants of the next stage. How is it possible?
- The influence of melanin is enormous given the astonishing precision with which the water molecule dissociates. And the energy that determines each of the various stages of formation of the different compounds, from simple to very complex, is the same (H_2 and e^-).
- And all the above should occur in an appropriate geochemical and geophysical environment.
- The effect of melanin on their environment is very powerful, not so much by the amount of energy but by the amazing consistency that shows. It is a kind of driven synthesis (melanin) which left very little to chance.

CONCLUSION

The importance of melanin as a source of energy, has only just begun to see. It is a disruptive knowledge that has applications in many branches of human knowledge. Having understood that way organisms absorb and process energy from the environment (of light), it will mean a great step forward, because we understand the arose and biology of life in a more concrete way.

So powerful is the way in which the melanin is going to change our very rooted and so wrong concepts about life, we think that the title of the book describes very closely the future:

Melanin, the Master Molecule.

This work was supported by Human Photosynthesis® Research Center.

REFERENCES

[1] Hill, H.Z. The function of melanin or six blind people examine an elephant. *BioEssays,* **1992**, *14*(1), 49-56.
[http://dx.doi.org/10.1002/bies.950140111] [PMID: 1546980]

[2] Larsson, P.; Larsson, B.S.; Tjälve, H. Binding of aflatoxin B1 to melanin. *Food Chem. Toxicol.,* **1988**, *26*(7), 579-586.
[http://dx.doi.org/10.1016/0278-6915(88)90228-1] [PMID: 3141255]

[3] Meredith, P.; Tadeusz, S. The physical and chemical properties of eumelanin. *Pigment Cell Res.*, **2006**, *19*, 572-594. Doi: 10.1111/j.1600-0749, 2006.00345.x.

[4] Sealy, R.C.; Hyde, J.S.; Felix, C.C.; Menon, I.A.; Prota, G.; Swartz, H.M.; Persad, S.; Haberman, H.F. Novel free radicals in synthetic and natural pheomelanins: distinction between dopa melanins and cysteinyldopa melanins by ESR spectroscopy. *Proc. Natl. Acad. Sci. USA,* **1982**, *79*(9), 2885-2889. b
[http://dx.doi.org/10.1073/pnas.79.9.2885] [PMID: 6283550]

[5] Chio, S.S.; Hyde, J.S.; Sealy, R.C. Temperature-dependent paramagnetism in melanin polymers. *Arch. Biochem. Biophys.,* **1980**, *199*(1), 133-139.
[http://dx.doi.org/10.1016/0003-9861(80)90265-9] [PMID: 6243905]

[6] Liu, Y.; Hong, L.; Bowers, C.; Wakamatsu, K.; Ito, S.; Simon, J.D. Comparison of the structural, chemical and spectroscopic properties of human black and red hair melanosomes. *Photochem. Photobiol.,* **2005**, *81*, 135-144.
[http://dx.doi.org/10.1562/2004-08-03-RA-259.1] [PMID: 15504086]

[7] Hong, L.; Simon, J.D. Physical and chemical characterization of iris and choroid melanosomes isolated from newborn and mature cows. *Photochem. Photobiol.,* **2005**, *81*(3), 517-523.
[http://dx.doi.org/10.1562/2005-03-02-RA-453.1] [PMID: 15790301]

[8] Borovanský, J.; Elleder, M. Melanosome degradation: fact or fiction. *Pigment Cell Res.,* **2003**, *16*(3), 280-286.
[http://dx.doi.org/10.1034/j.1600-0749.2003.00040.x] [PMID: 12753402]

[9] Sarna, T.; Pilas, B.; Land, E.J.; Truscott, T.G. Interaction of radicals from water radiolysis with melanin. *Biochim. Biophys. Acta,* **1986**, *883*(1), 162-167.
[http://dx.doi.org/10.1016/0304-4165(86)90147-9] [PMID: 3015231]

[10] Zareba, M.; Szewczyk, G.; Sarna, T.; Hong, L.; Simon, J.D.; Henry, M.M.; Burke, J.M. Effects of photodegradation on the physical and antioxidant properties of melanosomes isolated from retinal pigment epithelium. *Photochem. Photobiol.,* **2006**, *82*(4), 1024-1029.
[http://dx.doi.org/10.1562/2006-03-08-RA-836] [PMID: 17205626]

[11] Chedekel, M.R.; Smith, S.K.; Post, P.W.; Pokora, A.; Vessell, D.L. Photodestruction of pheomelanin: role of oxygen. *Proc. Natl. Acad. Sci. USA,* **1978**, *75*(11), 5395-5399.
[http://dx.doi.org/10.1073/pnas.75.11.5395] [PMID: 281688]

[12] Sarna, T.; Burke, J.M.; Korytowski, W.; Rózanowska, M.; Skumatz, C.M.; Zareba, A.; Zareba, M. Loss of melanin from human RPE with aging: possible role of melanin photooxidation. *Exp. Eye Res.,* **2003**, *76*(1), 89-98.
[http://dx.doi.org/10.1016/S0014-4835(02)00247-6] [PMID: 12589778]

[13] Rees, J.L. The genetics of sun sensitivity in humans. *Am. J. Hum. Genet.,* **2004**, *75*(5), 739-751.
[http://dx.doi.org/10.1086/425285] [PMID: 15372380]

[14] Wlaschek, M.; Tantcheva-Poór, I.; Naderi, L.; Ma, W.; Schneider, L.A.; Razi-Wolf, Z.; Schüller, J.; Scharffetter-Kochanek, K. Solar UV irradiation and dermal photoaging. *J. Photochem. Photobiol. B,*

2001, *63*(1-3), 41-51.
[http://dx.doi.org/10.1016/S1011-1344(01)00201-9] [PMID: 11684450]

[15] Smit, N.P.; Vink, A.A.; Kolb, R.M.; Steenwinkel, M.J.; van den Berg, P.T.; van Nieuwpoort, F.; Roza, L.; Pavel, S. Melanin offers protection against induction of cyclobutane pyrimidine dimers and 6-4 photoproducts by UVB in cultured human melanocytes. *Photochem. Photobiol.,* **2001**, *74*(3), 424-430.
[http://dx.doi.org/10.1562/0031-8655(2001)074<0424:MOPAIO>2.0.CO;2] [PMID: 11594056]

[16] Nicolas, C.M.; Robman, L.D.; Tikellis, G.; Dimitrov, P.N.; Dowrick, A.; Guymer, R.H.; McCarty, C.A. Iris colour, ethnic origin and progression of age-related macular degeneration. *Clin. Experiment. Ophthalmol.,* **2003**, *31*(6), 465-469.
[http://dx.doi.org/10.1046/j.1442-9071.2003.00711.x] [PMID: 14641151]

[17] Riesz, J.; Gilmore, J.; Meredith, P. Quantitative photoluminescence of broad band absorbing melanins: a procedure to correct for inner filter and re-absorption effects. *Spectrochim. Acta A Mol. Biomol. Spectrosc.,* **2005**, *61*(9), 2153-2160.
[http://dx.doi.org/10.1016/j.saa.2004.08.014] [PMID: 15911405]

CHAPTER 1

Intrinsic Chemistry of Melanin: Potential Approaches for Translational Medicine

Arturo Solís-Herrera[1,*] and **Sergey Suchkov**[2]

[1] *Human Photosynthesis Study Center, Av. Aguascalientes Norte 607, Pulgas Pandas Sur, Aguascalientes, México*

[2] *I.M. Sechenov Firts Moscow State Medical University and A.I. Konstantinov Moscow State Medical & Dental University, Moscow, Russia*

Abstract: Although the cell has been studied carefully, but its complexity exceeds the capacity for abstraction. However, it is believed that the findings about the intrinsic property of melanin to transform light into chemical energy through dissociating and re-forming the water molecule, which has not been previously known, marks a before and after in the knowledge regarding the cell-functioning. Thus, this current knowledge redirect towards a more effective way of practicing medicine that involves both cure and prevention.

Keywords: Energy, Glucose, Hydrogen, Melanin, Oxygen, Photosynthesis, Water dissociation.

INTRODUCTION

Living cells perform large, elaborate, in sync, and very specific tasks in order to maintain their highly elaborated biological functions. To carry out these highly complex functions, chemical gradients are required which work as an input of information, as well as signals to be sent into the cell from membrane receptors, and which can be equalized to the information processing; those which can also work as cellular responses, including reorganization of the actin cytoskeleton for motility and mitosis and also as the feedback of the underlying memory (genes).

Cellular biology has a fundamental requirement of a steady supply of chemical energy that has to be converted efficiently within very narrow ranges into useful work or capacity to produce various specific and general changes (since the energy needs of the cell are constant and incessant, in order to avoid the des-

* **Corresponding author Arturo Solís-Herrera:** Human Photosynthesis Study Center, Av. Aguascalientes Norte 607, Pulgas Pandas Sur, Aguascalientes, México; C.P. 20138; Tel/Fax: +524492517232; E-mail: comagua2000@yahoo.com

truction of the cell. It is increasingly clear that the generation and distribution of chemical energy are the determining factors in the form and function of organisms. Organisms function different from machines because they are naturally close to efficient cause which means that the catalysts needed for its operation must be generated internally. Thus, while organisms are complex, not all complex substances are considered organisms as for instance, melanin.

The organization in the space-time of the chemical reactions which makes up cellular processes and eventually integrates into a whole that we call life, has been exhaustively studied and so far the efforts are being made to understand the mysterious harmony that exists within the cell.

Thanks to the space program, now it is apparent that the fundamental, theoretic, and conceptual basis of embryology cannot be developed just by employing a traditional, molecular genetic framework, and that there is something else to it. It is now known that microtubules, very important cytoskeleton proteins, are differentially sensitive to gravity [1], which is a kind of energy. In addition to gravity modulation, other physical forces such as tension forces due to intercellular adhesion, the mechanical forces exerted during cell division, and physical waves of cytoskeleton expansion and contraction that traverse embryos [2], all provide significant physicochemical contributions to morphogenesis of embryos. The cell differentiation leading to morphological distinctiveness between cells in tissues and organs is further another example of self-organization or entropy reduction.

Physical forces, tension forces, intercellular adhesion and mechanical force are examples of energy that can be defined as "anything that produces a change". Thereby, the synthesis of new cells and the maintenance of existing cellular functions, both require an expenditure of energy.

Thermodynamics

The science that studies the passage of energy from one system to another and the transformation of energy from one form to another is called **thermodynamics**. Traditional thermodynamics is the statistical science in which observations are made on macroscopic samples.

Heat is considered the most common form of energy introduced into or released from chemical reactions, however it may involve light, electrical current, sound, pressure, magnetism, charge separation, *etc.* Since energy can manifest itself in various complex ways. Also, when gases are produced in a reaction, the work carried out on the surroundings by volume expansion becomes part of the energy balance, and conversely; reactions that consume a gas, undergo an opposite volume change, and work again is carried out on the system by the surroundings.

Laws of Thermodynamics

Thermodynamics is a derivative of two Greek words: *therme* ("heat") and *dunamis* ("power"). Thereby, it is the science that studies the power or energy contained in the heat, and also its conversion to other forms of energy. The term "energy" is itself derived from the Greek word *energeia* ("working"), the capacity to carry out a work. Something which has energy, has the capacity to carry out work, to exert a force through a distance, and to produce a change, whatever it may be.

Everything that exists in the universe is some form of energy, and everything that happens is also some form of energy conversion. Thus, the laws which govern energy generation, distribution and energy conversion are of paramount importance in understanding the world in which we live, and of course in understanding the cell itself.

The First Law of Thermodynamics states "Energy can be transferred from one place to another, or transformed from one form to another, but it can neither be created nor destroyed". This law is considered the most powerful and most fundamental generalization about the universe that scientists have ever been able to make. Recall that no one knows why energy is conserved.

Entropy, also known as the Second Law of Thermodynamics, describes a universally deteriorating order, the unavailability of a system´s thermal energy for conversion into mechanical work, is often interpreted as the degree of disorder or randomness in the system and also provides a measure of the amount of thermal or other kinds of energy that cannot be used to do a useful work inside a system.

The maximal level of this gradual decline into disorder, of this lack or order or predictability, is termed as equilibrium. There is a possibility of a number of molecular configurations to exponentially increase in a system with isolated equilibrium than in an unbalanced system. Therefore, in other words, the maximum degree of entropy is called thermodynamic equilibrium.

A thermodynamic system is said to be in thermodynamic equilibrium when it is in thermal equilibrium, mechanical equilibrium, radiative equilibrium, electromagnetically equilibrium and physics-chemical equilibrium. In a state of thermodynamic equilibrium, there are no net flows of mass, or of energy, no phase changes, and no driving forces (unbalanced potentials), within the system [3].

It is widely accepted that a system that is in thermodynamic equilibrium (TE) experiences no changes or modifications as it is isolated (to both mass and radiations) from its surroundings. It is not necessary for all the components to enter at the same time in TE, and often temperature gradients are present. Since 1851, William Thompson claimed that a thermodynamic system in equilibrium could be considered as inanimate.

The true thermodynamic equilibrium is rare, therefore local thermodynamic equilibrium can be considered, where one of the components of a system, for example, the gaseous component may be in dynamic equilibrium but only in certain parts or portions.

In any natural process, there exists an inherent tendency for the dissipation of useful energy. However, entropy could be considered as a thermodynamic process potentially reversible under certain circumstances.

Thereby, evolution is thermodynamically impossible. Evolution and entropy are opposing and mutually exclusive concepts [4].

The law of increasing entropy, so far; is an impenetrable barrier which no evolutionary mechanism yet suggested has ever been able to overcome. Entropy, derived from Greek *en* (meaning "in") and *trope* (meaning "turning") imply inward and downward. On the contrary, evolution, originated from a Latin word, means "out-rolling".

There is a general natural tendency of all observed systems to move from order to disorder, reflecting dissipation of energy available for future transformation, as is the law of increasing entropy [5].

Entropy and evolution are considered universal principles but are mutually contradictory. According to Sir Julian Huxley, the term evolution in the extended sense can be defined as a directional and essentially irreversible process occurring in time in which the basic or fundamental circumstances are maintained within certain ranges, in which its course gives rise to an increase of a variety and an increasingly high level of organization in its products. The whole reality is evolution, a single process of self-transformation [6].

Thus, on one hand, "all observed systems, apparently, move from order to disorder" and on the other, "the whole of reality gives rise to an increasingly high level of organization in its products", which means not only energy but information also, for instance, is a binary code.

The Evolutionary Cell Biology in Terms of Energy Conservation

Entropy can be expressed mathematically in terms of the total irreversible flow of heat. The available energy has to flow from a higher level to a lower level. Eventually, when a lower level is reached, energy is still in existence, but seems no longer capable of doing a perceptible work.

The evolution in biology is built on innovation based on the framework of energy conservation. Evolution is characterized by directional processes of change, which are constant and universal

Inside the cell, as in universe, no process can be 100% efficient, which is all of the available energy cannot be solely converted into work. Some energy will be deployed to overcome friction, and will be degraded to heat energy; therefore, there is a gradual loss in the availability of energy for the future performance of work. This suggests that the evolution is a process that eventually would stop due to energy loss.

Billions of years of continuous evolution have shown that cellular mechanisms exist which overcome the Second Law compensating efficiently the energy losses, and since the evolution is characterized by generating progressive and complexes changes in the cell, therefore the organisms are distinguished by decreased entropy.

This apparent paradox between evolution and entropy has tried to explain with arguments such as "the universal Second Law of Thermodynamics shows, apparently; that things become more disordered through time, not more complex, as evolution insists" [7].

Therefore, the biological evolution that converts unicellular organisms to multicellular, *i.e.* humans, with and obvious increase of order and complexity would violate the Second Law. Thereby, living things make unusual things happen.

The most devastating and conclusive argument against evolution is the entropy principle, unless some as codified energy could exist. The Second Law of Thermodynamics implies that evolution in vertical sense is completely impossible. The law of increasing entropy is a universal law of decreasing complexity, unless something plays in opposed way, whereas evolution is supposed to be a universal law of increasing complexity [8], but it is necessary to declare something as codified energy.

Evolution requires some universal principle, as energy codified; which increases order, causing random particles to eventually organize themselves into complex, non-living systems to become living cells (chemical evolution, thereby chemical energy). But this universal principle would have to exist before the life, and after that the problem may arise as how to concatenate the universal principle with the biological mechanisms that coordinate cell function and do unusual things which happen within organisms.

The Cell, the Forces of Gravity and other Physical Forces

Thanks to the space program, now it is apparent that the fundamental conceptual basis of embryology cannot be developed just by employing a traditional or classical molecular genetic framework. It is now known that microtubules, very important cytoskeleton proteins, are differentially sensitive to gravity [9]. In addition to gravity pull, other physical forces such as tension forces due to intercellular adhesion, the mechanical forces exerted during cell division, and physical waves of cytoskeleton expansion and contraction that traverse embryos [10], all provide significant physicochemical contributions to morphogenesis of embryos.

The cell differentiation leading to morphological distinctiveness between cells in tissues and organs are further examples of self-organization or entropy reduction. However, the second law of thermodynamics states (theoretically) that the entropy of an isolated system never decreases, because isolated (from mass and radiations) systems spontaneously evolve toward thermodynamic equilibrium, the state of maximum entropy.

The Cell, an Apparently Chaotic System

Chaos is defined as the study of those systems with the property that small change in the initial conditions can lead to very large effects in the subsequent evolution of the system. Thereby, chaotic systems are inherently unpredictable [11].

The weather is a clear example: small changes in the temperature and pressure over the ocean can lead to large variations in the future development of a storm system. If it is applied to cell biology, small changes in some variables as temperature and pressure can lead to significant and unpredictable changes in cell function and therefore in health.

The cell is the maximum known example of reducibility able to express life. Since it is the result of the patient work of nature over four billion years of evolution, it is incomprehensible to us. But within this apparent chaos, there is actually a strict order within certain narrow ranges.

The hundreds or thousands of biochemical events that happen constantly, minute by minute, hour after hour, day and night inside the tiny cell, do not occur at random in any way. Forming something similar to what we call continuous random variables that are processes that can take on any real-number value within a certain range and time. They are characterized by a density function curve such that the area under the curve between two numbers represents the probability that the determined random variable will be between those two numbers.

Therefore, chaotic systems and cells can exhibit certain kinds of regularities. Thus, the apparent chaos regarding the order and timing of chemical reactions that occur intracellularly shows those 4 billion years which far outstrip our ability to abstraction.

The Cell from the Point of View of the Entropy

At the macroscopic level, living systems are thermodynamically open (characteristically tends to decrease disorder, disorganization) and far-from-equilibrium systems [12]. Hence the balance of entropy at this level must necessarily involve chemical energy used to maintain cell metabolism, as well as heat (disordered energy) and waste (by-products) product dissipation into the external environment, all the while being sensitive to physical and chemical forces acting on it from the environment.

Living cells are heat-dissipative, open systems, and far-from-equilibrium that constantly tend to decrease the entropy utilizing an influx of chemical energy and codified information carried by it, besides molecular material in a multi-compartment structure with specific functional characteristics. Therefore, apart from the dissipation of heat, there are also very orderly processes of expulsion of waste products from molecular material into the environment.

Entropy reduction was analysed in the middle of the last century by Schrödinger [13], whose equation describes how the quantum state of a system, interpreted, for instance, as the probability of a particle being detected at a certain location evolves over time.

Thus, it is fundamental to understand the entropy reduction, which is only possible in an open entropic system, such as universe, the cell and melanin, and how all of these tend to decrease the disorder as a constant and astonishing feature, and the cell contributions arising from internal self-organization (as in melanin), information storage and transfer, at a single cell level as the only way to reconcile this with laws of thermodynamic is by balancing orderly and congruously, these free energy changes with metabolic energy expenditures.

Thus, extraordinary observations that can be very useful for the advancement of medicine, as apparent relationship between microtubules and rearrangements of cytoplasmic components induced by pre-cleavage waves [14], have been in the vacuum, since the correlation between these events and processes that induce and govern them, are unknown, only we are able to perceive them as periodic calcium waves crossing the egg once it is fertilized.

Interestingly, pigmented cells contain high amounts of calcium, reflecting the enormous calcium-binding capacity of melanin. No tissue in the entire body of the embryo or adult exceeds the calcium binding capacity of pigmented tissues [15]. For instance, bone which has rather lower binding capacity despite its calcium concentration being one thousand times that of pigmented tissues. However, calcium is not bound but is deposited in mineral form [16].

Thus, in the interaction between energy and mass that occurs within the cell is where the secret of health lies, and thus of life itself. It is not possible to cheat in energy cell processes. Glucose beguiled molecular biology as source of energy.

Melanin, the Fundamental Molecule of Life and Evolution

Despite the unique and complex physicochemical properties that have been described about the melanin molecule, in biology, melanin pigments serve as the major determinants of the skin color and also as the major sources of skin protector against radiation, also preventing sun-induced skin damage as well as skin cancer development [17].

The role of melanin is the same, regardless of its location and function, as derived from its physical and chemical properties.

Photo-protective Role of Melanin

With the exception of the skin, in which melanin is generally accepted to play a photo-protective role [18]; the biological role of melanin in other body places, like ocular tissues is still subjected to intense debate [19]. The coloration of human skin is highly variable. This variation so far, is considered primarily hereditary, but environmental factors certainly have an effect.

If the property of melanin is taken into account, as previously unknown, to transform light energy into chemical free energy, which, coupled with the fact that it is the same process in each and every cell in the body, then the pigmentation skin takes on another meaning, as for that the amount of melanin in the skin regulates the amount of light that should reach the cells within the body (Figs. **1-3**), which also requires visible and invisible light to perform its functions.

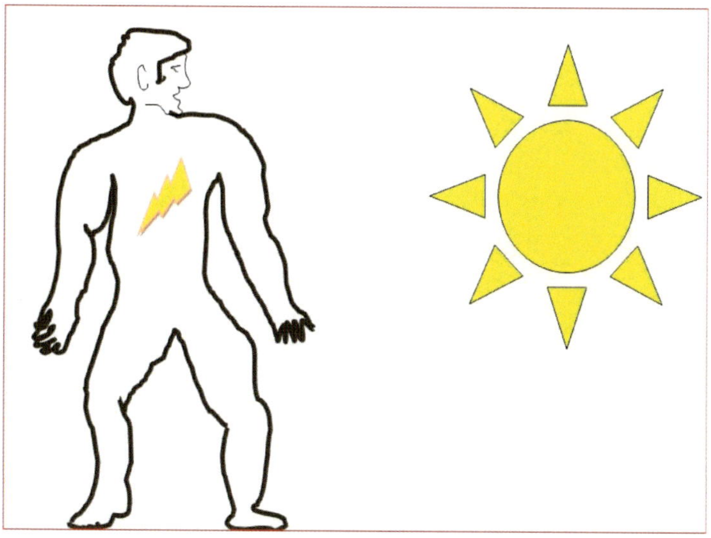

Fig. (1). When the solar radiation is intense, such as in the tropics, the amount of melanin in the skin is raised in order to maintain the levels of light that enter into the body, as these must be within certain ranges ($E = M^{3/4}$), that is, must not be too much nor insufficient.

Fig. (2). In Caucasian regions where the amount of light is lower, the amount of melanin in the skin decreases to allow passage of an increased amount of sunlight, so that the cells inside the body can get the right amount of energy light and consequently chemical energy that allows them to efficiently carry out their duties ($E = M^{3/4}$).

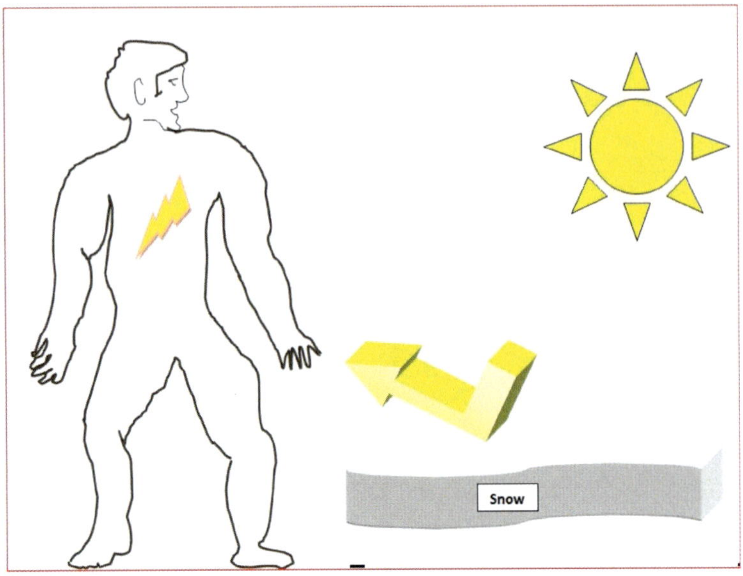

Fig. (3). The dark color of hair, eyes and skin from the Eskimo population is explained by the albedo, which is defined as the reflection of light on any surface, therefore, although the amount of solar irradiation is even lower than in Caucasian regions, pigmentation is greater as the reflection of sunlight on snow increases the amount of light energy available.

The process of transformation of light energy into chemical energy is the same in all living things, and evolving to adapt to the conditions of the environment, for example, melanin in the trees is located in the trunk [20], called as lignin. And in the trunk of the trees, the concentration of melanin in the cortex also reflects the intensity of the sunlight, as well as the coloration of the skin in humans (Figs. **4** and **5**).

Melanin and Other Pigments in Biology

Pigment is defined as the natural coloring matter of animal or plant tissue. These are of 5 types present in living things: hemoglobin, chlorophylls, carothenoids, flavanoids and melanins. The prosthetic groups of the first four substances are orderly arrays of single and double bonds, except melanin. Their protein moieties (first four substances) are easily studied using conventional techniques, except melanin.

Of all the pigments, the most widely distributed in nature, is melanin, since it is practical on life forms covering the various biological kingdoms (Figs. **6**, **7**). However, paradoxically it is the most neglected.

Fig. (4). Photo-system in plant is also comprised of light/melanin/water, but the melanin in the trees is in the trunk, and the amount of melanin present in the cortex, which in plants is called lignin, also depends on environmental factors, such as the amount of light present, and in bright light, for example, in tropical areas, the amount of melanin (lignin) is increased which in turn decreases the amount of light reaching the interior of the plant.

Fig. (5). In low-light areas, the bark of the trees is paler, as the amount of exposure of light to melanin (lignin) is smaller, which allows the amount of light energy to reach the interior of the plant in greater amount in order to fit appropriate to the metabolic needs of the plant.

Kleiber law $E = M^{3/4}$ is satisfied in all biological kingdoms.

Fig. (6). Worm, a non-arthropod invertebrate, insect larva; stretch marks correspond to dark melanin molecule.

Fig. (7). Black butterfly, Common Brangas, Brangas Neora (Hewitson 1867); family: Lycaenidae, subfamily: Theclinae. The dark color is the result of the presence of melanin.

Apart from a few very notable exceptions, the materials, structural and chemical physics have been largely absent from the melanin debate [21]. With recent advances in theory and computational methods, quantum and condensed matter physicists are now becoming interested in apparently intractable problems such as melanin. So far, photo-protective action is commonly believed to be its main biological role (Sarna, 1992; Sarna and Plonka, 2005; Sarna and Swartz, 1998).

Despite many decades of work within the organic chemistry, biophysics and pigment cell communities, still there is no complete picture of the macromolecular structure of eumelanin pigment, the most studied of all classes of melanin and the predominant type of the melanin pigment in the eye; either in its protein-bound natural state or isolated synthetic form.

The ultimate goal of melanin research is to get a full mapping of how molecular and cellular scale structures relate to macroscopically observable properties and functions, as considered utopian till now. It is duly said that melanin defines the characterization and analysis through current methodology [22].

Melanin Functions Described in the Literature (Table 1)

In the literature, the following references to the functions of the melanin molecule can be found in biology.

Table 1. Over time, they have been describing in different sources of information, the supposed biological functions of melanin in the organism as well as its physicochemical properties, all of them interesting, and whose explanation was little understood, but not It had detected its main function which is to transform visible and invisible light into chemical energy through the dissociation of the water molecule, such as chlorophyll in plants.

Function	*Explanation*	*Confusion due to*
Camouflage	Avoid predators	Not a very good
Adornment	Sex appeal	Not a very good
Sunscreen	SPF 2	Not a very good
Radical Scavenger	Melanins are unique among biological molecules in that they continuously emit a free radical signal [23]	Not a very good
Photo- and Radio-sensitizer	Produce free radicals and damage cellular macromolecules including DNA.	Free radicals escape easily.
Binds to drugs	Detoxify xenobiotics	Drug-melanin interactions may be a two-edged sword.
Energy Transducer	Transform different types of energy into heat and by this means dissipate them [24].	If the energy input is too great, the capacity of the pigment to detoxify the radicals is exceeded and potentially damaging species (hydroxyl radicals OH) are produced [25].

(Table 1) contd.....

Function	*Explanation*	*Confusion due to*
People with dark skin is more prone to debilitating skin ailments [26]	This hypothesis relies in part on frost bite studies from Korean and earlier wars.	Cold treatments of black and white areas of guinea pig skin demonstrate a greater sensitivity of the dark skin.
Skin cancer hypothesis	Dark skin protects against skin cancer	Melanin-mediated radical production is potentially lethal to cells [27].
Vitamin D hypothesis	Dark skin in low sunlight prevents activation of vitamin D precursors in the skin	Racial pigmentation does not interfere with normal production of Vitamin D active forms [28].
Selective advantage of the loss of pigment in northern climates	Is not all clear	Eskimos do not have light colored skin.
Melanin can absorbs photons, electrons, noise, free radicals and active oxygen species	Melanin can be overwhelmed and enhance the yield of free radicals.	It now becomes a generator of hydroxyl radical, the most evil of the active oxygen species.
Melanin has the capacity to act as a pseudo-super-oxide dismutase and can convert superoxide anion radical to hydrogen peroxide.	Antioxidant property	Under certain circumstances melanin is pro-oxidant by itself.
Melanin binds metals	Protect versus poisoning with metals	Once melanin binding capacity is overwhelmed, and then becomes a reservoir of toxic substances.
Melanin is able to act like amorphous semiconductors [29]	Its relaxation times are slow, a characteristic of amorphous semiconductors [30]	Amorphous semiconductors differ from crystalline semiconductors that the energy of quanta absorbed by different domains of the molecule may be different and therefore cannot be evaluated.
Melanins are important in the inner ear [31]	Albinos frequently suffer from hearing loss and is inversely related to skin pigment type [32]	Melanins may convert acoustic energy into heat [33]
Melanin can transform light energy into electrical energy and into heat [34]	Melanin can also store electrical energy and convert it slowly into heat.	Melanin's conductivity increases as temperature increases in contrast to metals in which conductivity decreases with increasing-temperature.

It is known in a more or less detailed structure-activity relationship of biological pigments, except for melanin. The formula of melanin is entirely theoretical due to the fact that characterization and analysis methods to date are not sufficient to unravel the mysteries about it.

EPR (Electron Paramagnetic Resonance) studies have provided evidence for the presence of semiquinone radicals ($QH^{-\cdot}$ /$Q^{-\cdot}$) In eumelanin. It is not easy to define whether these free radicals present within the melanin have cytotoxicity capabilities in more or less degree, and under what conditions [35].

In spite of its high molecular weight, melanin added to the cell culture medium is taken up by the cell end and into the cytosol and other subcellular compartments.

Melanin is produced in a number of areas including the brain, eyes, adrenal gland, and hair. Production of melanin in melanocytes results in the pigmentation of skin. The production of melanin (melanogenesis) is a complex cellular process, involving over 100 genes, which regulates, among many other things and involves melanin biosynthesis and intracellular trafficking of melanogenic enzymes to melanosomes.

The location of melanin in obscure internal body sites, their exotic chemical and physical properties, the sequestering of the pigment in protective subcellular particles, and diseases melanin-related, all lead to speculation that melanin have other functions than simple solar protection.

Melanin as an Energy Source

The fact that melanin overnight becomes the main actor of cell biology is something that breaks many paradigms believed impassably. However, if the physical and chemical properties of the compound are analyzed along with its seemingly strange location within the cell and the body, the unexpected and surprising findings make a lot of sense.

Melanin pigments are found in most living organisms, including bacteria, plants, and animals [36], and are common elements of a variety (probably all) of vertebrate tissues [37], *e.g.* skin, nails, hair, eye, inner ear, substantia nigra, locus ceruleus, meninges; *etc.*

It is very frequent to find the claim that melanin is not a vital compound as organisms called albinos are considered viable, which is a disadvantage for it as compared with other pigmented bodies. But it may be a misunderstanding, because albinos also possess melanin, and what actually happens may not be in the right quantities (Figs. **6-9**).

Albinism is regarded as a reduced synthesis of melanin rather than its total absence [38]. A careful reading of the literature regarding albinism demonstrates an evolution in our understanding of this complex syndrome. A number of confusing terms such as “incomplete”, “partial”, or “imperfect” albinism were used. The identification of normally pigmented offspring arising from the mating of two individuals with oculo-cutaneous albinism provides proof through complementation that genetic heterogeneity exists [39]. The hair bulb test, in which a freshly plucked hair bulb was incubated in tyrosine or dihydroxyphenylalanine, separated oculo-cutaneous albinism, a surface somewhat; into "tyrosinase-negative" and "tyrosinase-negative". It is now apparent that the classification of oculo-cutaneous albinism based on hair bulb test or the clinical phenotype is inaccurate.

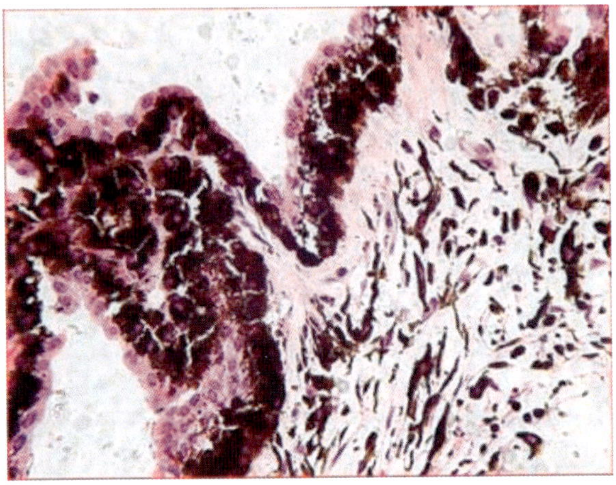

Fig. (7). Micro photograph of the eye of the albino Wistar rat, in the region of the ciliary processes, note the presence of abundant pigment (melanin).

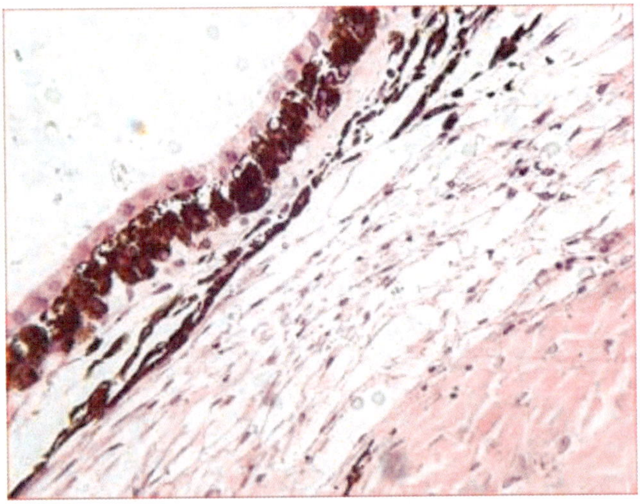

Fig. (8). Photomicrograph of a histological section of the Wistar albino rat eye, anatomical area corresponds to a region near the ora serrata, some muscle fibers were observed corresponding to the longitudinal portion of the ciliary muscle, and the left side of the photograph shows the retinal pigment epithelium, which in that area has an inner layer with less pigment and an outer portion with increased amount of melanin.

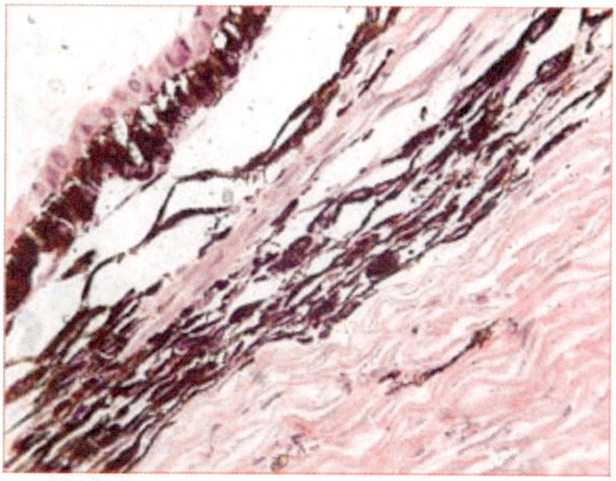

Fig. (9). Microscopically photographed area corresponds to an area midway between the ora serrata and the anterior chamber angle. The abundant amount of melanin in both retinal pigmented epithelium and in the muscle fibers which correspond to the ciliary muscle is easily discernible. The sample is from a Wistar albino rat eye.

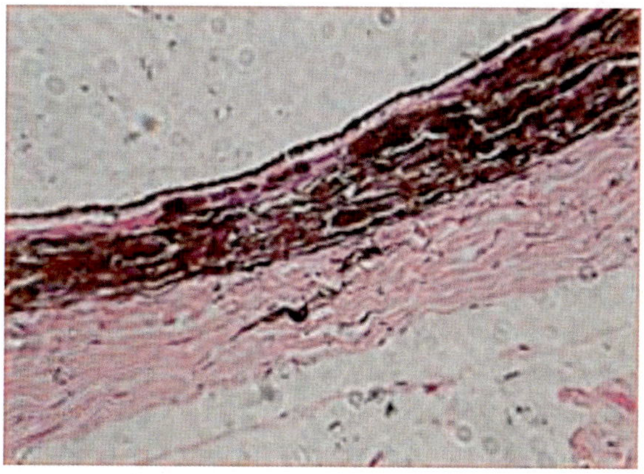

Fig. (10). In this section of the eye of albino Wistar rat, corresponding to a region near the macular area, it is appreciated that the choroidal layer thickness increases considerably. Choroid word means grape color due to high amount of melanin.

Currently, the term "albinism" refers to a group of genetic abnormalities of melanin synthesis associated with a normal number of melanocytes in the skin and the eyes [40]. Cutaneous and ocular hypopigmentations are not sufficient to define albinism. The precise definition of albinism includes ocular and cutaneous hypopigmentations associated with specific changes in the development and function of the eyes and the optic nerves [41]. Changes in the auditory system are found in the brain in albinism but are not included in clinical evaluation of a child or an adult with albinism [42].

The alterations in the amount of pigment that may give rise to functional alterations are also observed in mammals (Fig. **11**).

Fig. (11). Albino hedgehog, the amount of melanin is less than the native species, which decreases its ability to survive outside of captivity. Interestingly, domestication seems to decrease the amount of melanin.

Characteristic changes in the development and function of the eye and optic nerves are common to all types of albinism. So far, it has been hypothesized that these changes are related to the reduction in melanin in these tissues during development and postnatal life rather than to a genetic or pleiotropic effect of the responsible genes, although no explanation was offered in the slightest regard as to why the diminution in the synthesis of melanin had a remarkable and widespread effect. The above hypothesis has been proposed because the mice study with different pigmentation abnormalities and patterns has revealed at least 60 different loci that are involved in the processes of melanocytes maturation and pigment production [43]. At present, many pigmented-related genes have been isolated, and more are being reported each year, thereby, many different gene products are necessary for normal pigment formation. But the expression of genes, however complex and incomprehensible it may seem, has a common requirement of energy.

Moreover, since the tissue damage is widespread, and the least amount of energy available explains it well, can be considered indirect evidence for the finding.

The Chemistry of Melanin

The history of melanin research is usually long and entangled [44]. The reviewer had faced the problem of recording theories, assumptions, and speculations that have often become axiomatic, rather than well-rationalized experimental facts.

Among the relevant historical data, we have the following:

In 1765, Le Cat first extracted the skin of a black Ethiop and found that the isolated pigment was chemically similar to those of the choroid of the eye and, notably, squid ink.

In 1896, Abel and Davies extracted the pigment from the entire skin of a Negro cadaver and found that it contained carbon, hydrogen, oxygen, nitrogen and some sulfur.

In 1927, Raper elucidated the early stages of the oxidation of tyrosine to melanin to tyrosinase. He correctly identified the red intermediate as dopachrome and succeeded in isolating 5, 6- dihydroxyindole (DHI) and 5, 6-dihydroxyindole-2-carboxilic acid (DHICA) as methyl ethers.

In 1949, Lerner, Fitzpatrick and coworkers provided evidence that melanogenesis in mammals' proceeds, as in lower organisms, *via* the action of tyrosinase on tyrosine suggesting that it is a characteristic highly conserved during evolution and therefore important for the nature.

When performed *in vitro*, the oxidation of tyrosine with tyrosinase leads invariably to the formation of a black pigment. By contrast, tyrosine metabolism in mammalian melanocytes results in a considerable broad palette of colorations (numerous shades of yellow, red, brown, black, and black of mammalian hair).

Apart from genes, other factors are also found regulating melanogenesis. The discovery of cysteine as the major determinant of the switching between the eumelanin (black) and pheomelanin (reddish) pathways provided the first evidence for the existence of regulatory mechanisms acting distal to tyrosinase. Recently, a number of other regulatory factors have been described that affect melanogenesis at different levels and with different modalities such as glutathione and related enzymes (reductase, peroxidase), tyrosinase-related proteins (TRP1, TRP-2), catalase [45], superoxide dismutase [46], and thioredoxin reductase [47].

The Physical Properties of Melanin

The melanins are usually heterogeneous compounds with quite intractable physical properties that resist attempts to characterize them by simple physical-chemical approaches [48]. Thereby, attempts to characterize melanins have involved the use of a great number of different and often complex techniques. These techniques have led to large amounts of data, but the nature of the data obtained from many different techniques, combined with the heterogeneity of the melanins, has resulted in incomplete and sometimes contradictory conclusions.

In most cells, melanin is synthesized in a specific organelle: the melanosomes. Depending on its location, melanin has some peculiarities, for example, in the skin, after fusion with lysosome; the epidermal melanosomes are degraded and broken into "melanin dust" [49]. It has been reported that neuro-melanin accumulates with age in the substantia nigra of human brains until it reaches a maximum level at the age of 50-60, and then it gradually decreases [50].

Optical Properties

The absorption of light is one of the most obvious and important properties of melanin. This has been proven to be an immensely complex topic, in terms of both the understanding of the mechanisms and consequences of optical absorption by melanin. Melanin thrives on pressure, as the more energy is present, the more is absorbed.

Observable optical properties are a complex function of melanin´s ability to mostly absorb, scatter, and reflect (if any) light at different wavelengths. Melanin is the substance that is known as obscure. Melanin has a very efficient non-radiative de-excitation system, which is "unexplained" and dramatically improves when hydrated (Fig. **12**).

Indirect Evidence of the Unsuspected Property of Melanin

Direct demonstration of the intrinsic property of melanin to dissociate and re-form the water molecule is a formidable challenge, otherwise has been detected years, decades or centuries ago. It is not possible to directly prove the phenomenon of water dissociation, for example, watching through a microscope, because hydrogen is the smallest atom and the reaction occurs in a nanosecond range; but can be demonstrated by indirect evidence, like many other things in all areas of science, as there is a classic example of the Pythagorean theorem, which is defined as the square of the length of the hypotenuse of a right angle is equal to the sum of the squares of the lengths of the other two sides.

Till today, the theorem of Pythagoras remains the most important single theorem in mathematics. It is not known who discovered it and early indications about it date back to 1800 BC, in the Babylonian cuneiform tablet known as Plimpton 322; and it cannot be proven directly.

Fig. (12). The melanin in aqueous solution and solid state.

As indirect evidence of the intrinsic capacity of melanin to split and re-form the water molecule, there is a peculiar characteristic of melanin as being the only substance that both *in vivo* and *in vitro* have free radicals. This has long been known [51] and has engendered a very intense research activity. ESR signals of eumelanins are apparently provided by semiquinone units deeply embedded into the pigment granules.

The presence of such semiquinone units is compatible with the frequently cited ability of eumelanins to act as redox pigments with both oxidizing and reducing properties toward oxygen radicals and a range of chemical redox systems [52].

In order for such electron-exchange reactions to be considered as reflecting true redox properties, it is essential that they are reversible; that is, a given pigment sample should be able to cyclically accept and donate electrons in the same amount without structural modifications. Therefore, the finding about the intrinsic property of melanin to split and re-form the water molecule is compatible with the above statement.

Melanin consumes oxygen to an extent depending on the specific structural features and reducing capacity of the pigment. This consumption can be explained because melanin uses absorbed oxygen to form water.

Racial Differences in Ocular Oxygen Levels

Comparing PO_2 in different regions in the eyes of patients undergoing intraocular surgery, Siegfried *et al.* [53] found that the PO_2 value was significantly higher in African-Americans patients at all 5 locations compared with Caucasian patients. Racial differences in oxygen levels in the human eye reflect an important difference in oxygen metabolism. This is indirect evidence that the greater amount of melanin will capture greater light energy and will thus dissipate in the same amount through dissociation and re-formed water molecule.

Kleiber Law ($E = M^{3/4}$)

Observers of living organisms since Galileo have recognized that metabolic activities must somehow be limited by surface areas than body volumes. Kleiber [54], since 1932, noticed that when rate of heat production is plotted against body weight on logarithmic scales for animals over a size range from rats to steers, the points fall extremely close to straight line with slope 0.75, which is a very interesting and constant fact, but remains unexplained to this day.

The results have been confirmed for animals as different in size as the mouse and the elephant, and have been verified for other metabolically related variables, such as rate of oxygen consumption. As it is considered so far, biological laws are not derivable from physical laws in any simple sense. But Kleiber's law is fulfilled in all living things, from bacteria to the plant and animal kingdom. In addition to this law, which can be expressed as $E = M^{3/4}$ seems to govern the size and shape of all living things, as apparently is true in every one of them. This, in addition to be an astonishing fact, cannot be explained in the light of current knowledge, although it was first described in the mid-1800s.

The statement of Kleiber´s Law can be applied to the study of the ratio of many physiological periods to one another and is found to be nearly constant, independent of scale. Thus the ratio of gut pulsation time to pulse time is nearly the same in all mammals, and each animal lives for approximately the same number of heartbeats or breath cycles [55]. Thereby, it seems that there is "something" like a universal mechanism which regulates the metabolic rate of cells in general, no matter which species belong to it, because it appears to apply to both the animal and the vegetable kingdom, and to bacteria also.

Kleiber [56] wrote: *"When the concepts concerned with the relation of body size and metabolic rate are clarified, then comparative physiology of metabolism will be of great help in solving one of the most intricate and interesting problems in Biology, namely the regulation of the rate of cell metabolism."*

Any comparison between different species presupposes a common basis, and so far, attempts to discover something of a unifying principle,or the great regulator in metabolism have been futile.

The metabolic rate (heat production per unit of time), in particular the basal metabolic rate of humans, is generally expressed in kilocalories per square meter of body surface. This procedure is based on the theory that in animals of different body size, the metabolic rate is proportional to their respective surface areas, a theory called surface law. This dates from 1839, when Sarrus, mathematical, and Rameaux, MD, presented the thesis at the Royal Academy of France, in July of that year.

And one way to summarize it would be, 'the rate of heat production of large and small animals is in proportion to their respective surface areas or the 2/3 power of their body weights.'

Why Big Animals Use Energy More Efficiently Than Little Ones?

This is a question that fascinated biologists a long time ago. An elephant, for example, is 200,000 times heavier than a mouse, but uses only about 10,000 more energies in the form of calories it consumes. The bigger you are, you actually need less energy per gram of tissue to stay alive [57]. This is an amazing fact. This relationship between the mass and energy use of any living thing is governed by a strict mathematical formula:

$$\mathbf{E = M^{3/4}}$$

That is universal or almost universal across life. So it operates from the tinniest bacteria to whales and Sequoia trees. It had been implemented in the 1930s, and no one has been able to explain it. According to the Kleiber's law, the mechanism that regulates the metabolic rate of eukaryotic cells (and prokaryotes possibly also) seems to be the same in all living beings therefore it can be considered as universal and unique. Unfortunately, and despite the best efforts, no one has been able to find something like a unifying principle or the great regulator that could be applied so widely in biology as Kleiber's Law itself.

However, the intrinsic capacity of melanin to split and re-form the water molecule, a previously unknown fact, may be the long-sought explanation to Kleiber's law. Since melanin is also found in all biological kingdoms, it is thought that its role would be similar in all of them; to transform light energy into chemical energy freely through the dissociation of the water molecule. Thereby, it is a useful answer in comparisons of living things on both the microscopic and the gross scale, as part of the growing science of form, which asks precisely how organisms are diverse and yet so alike.

If the free chemical energy is a determinant of each and every one of the chemical reactions that occur in the cell, more even in all cells, coupled with the observation that the bigger is the living entity the lesser food it needs (while larger the antenna, then better capture electromagnetic radiations), then the intrinsic melanin property to dissociate and re-form the water molecule is very consistent as rector basic process, as it generates all of the free chemical energy that the cell needs daily and constantly, and on the other hand, the bigger the living entity is, the best it serves as an antenna, *i.e.* it better captures electromagnetic radiations.

Melanin Functions in the Eye

The literature on the functions of melanin in the eye has not changed much since we started the research in 1990. The concepts remain the same to date. Even though, melanin is not considered a visual pigment, honor conferred for decades upon Rhodopsin, a light-sensitive pigment that has peak sensitivity of about 510 nanometers, in the green part of the spectrum. Rhodopsin has a special property that when it absorbs a photon of light, it changes its molecular shape and at the same time releases energy, although not specified what type of energy. Thus as a result of a single pigment molecule being bleached, millions of pores close off. It had been very hard to imagine how the bleaching (from all-cis to all-trans) of a single molecule could lead to the closing of the millions of pores that the observed potential changes would require.

And the explanation more popular these days, which involves the cGMP and a cascade of events as theoretical as the previous, mostly because they are inferred from laboratory studies performed in other species, where the histological differences photoreceptors are relevant, but apparently no one has paid attention to the fact that there is a data that does not change, the omnipresence of melanin.

Melanin is always in close contact with the photoreceptors, whatever the species, whichever the anatomy of the photoreceptor. Taking into account the intrinsic property of melanin to dissociate and re-form the water molecule, it is easier to understand the reason of the insistence of nature in place as it is always in the same way in terms of its energy production and more specifically, free chemical energy. Therefore, the human being is not an exception, and the following description of the anatomy of the retina, taken from a textbook, verifies it:

"*Rods and cones are at the back of the retina; thereby the incoming light has to go through the other layers of retina itself in order to stimulate them. Presently, it is not fully understood why the retina develops in this curious backward fashion. Melanin mops up the light that has passed through the retina, keeping it from being reflected back and scattering around inside the eye; it has the same function as the black paint inside a camera [58]. The melanin-containing cells also help chemically restore the light-sensitive visual pigment in the receptors after it has been bleached by light. For both functions, the melanin pigment must be close to the receptors".*

I wonder, referring to the phrase *"also help restore chemically light-sensitive visual pigment",* since it refers to Rhodopsin, and does not explain more about this entirely theoretical chemical mechanism of restoration.

How the Results Have Been Achieved

The story begins in 1990, during an observational, descriptive study about the three leading causes of blindness, initiated by the reason that these are the same recurring diseases worldwide prevailing for 60 years or more: Glaucoma, diabetic retinopathy, and age-related macular degeneration. Given that the incidence and prevalence of these have not changed to date, we can infer that the treatments available, expensive, heroic, sophisticated or pompous are not working. So we decided to start a protocol in order to find new treatments that could alleviate the loss of vision of patients.

Our working hypothesis at the time was that if we magnify the optic nerve blood vessels in the living patient, whose diameter is 1200 microns (12 human hairs together) then maybe we can find early signs of disease, allowing us to institute early treatment. The above was based on the belief that any metabolic need of the tissues is completely filled by the blood vessels. There are 40 billion capillaries in the body that hold only 5% of total blood volume. At any given time, only a fraction (25%) of capillaries is fully filled with blood, especially in tissues at rest. Capillaries are essential for the delivery of oxygen to the tissues and the exchange of nutrients between blood and interstitial fluid surrounding the cells [59]. Capillaries form functional units known as capillary beds and these are not uniformly distributed among the different tissues. Sites of high metabolic activity (such as the liver and kidneys) contain numerous capillaries, while sites with little metabolic activity (such as the lens of the eye) are capillary-free.

So our main variable under study is the anatomical characteristics of the tiny blood vessels containing the optic nerve. The first challenge to overcome, this in 1990, was to use the new digital methods fundus photography, so that the capillaries of the optic nerve could be amplified in the clearest way possible, and this in the living patient, which fortunately no risk or harm involved to the patients examined, as fundus photography is a usual method of examination in ophthalmic patients, even until today. The difference then was attaching the new digital methods which allowed the use of different filters that allow us to see different layers of tissues, making a kind of optical dissection because the different structural layers of the optic nerve head and retina have selective affinities for the different wavelengths of the spectrum visible. Spectrophotometry is something like the lab, where monochromatic light using the intrinsic affinity of each molecule to specific wavelength allows us to recognize it, the same thing happening with the retina and optic nerve head, where the different structural elements *et al.* vessels and nerves, have different bright affinities, allowing us to get clear vision when using monochromatic light of different wavelengths to illuminate.

And in a relatively short time, we could achieve the appropriate magnifications (Figs. **12-14**), which together with the use of different existing filters; allowed us to observe the anatomical details we needed, but at the same time we begin to notice the omnipresence of melanin in almost 6000 patients examined (Figs. **15-17**). At three months into the study, melanin became the main study variable, relegating to the background the observation of blood vessels.

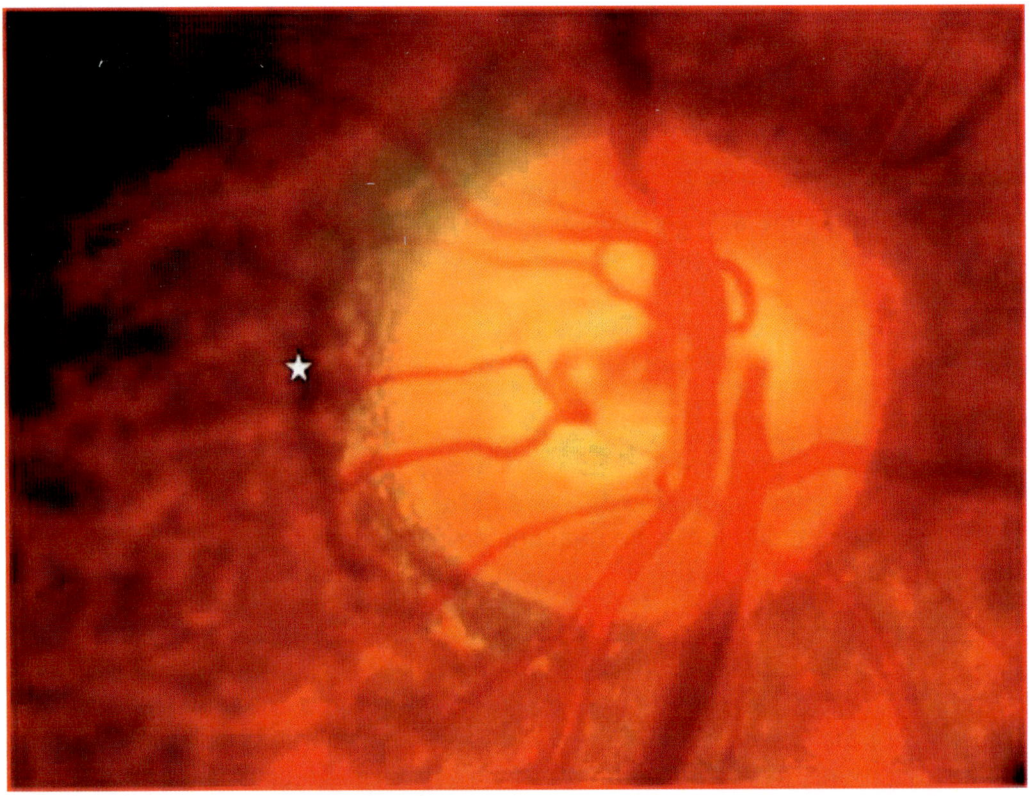

Fig. (13). The optic nerve in the patient's life, the diameter of 1200 microns is equivalent to twelve human hairs together. The white star shows the presence of melanin in the edge of the disc.

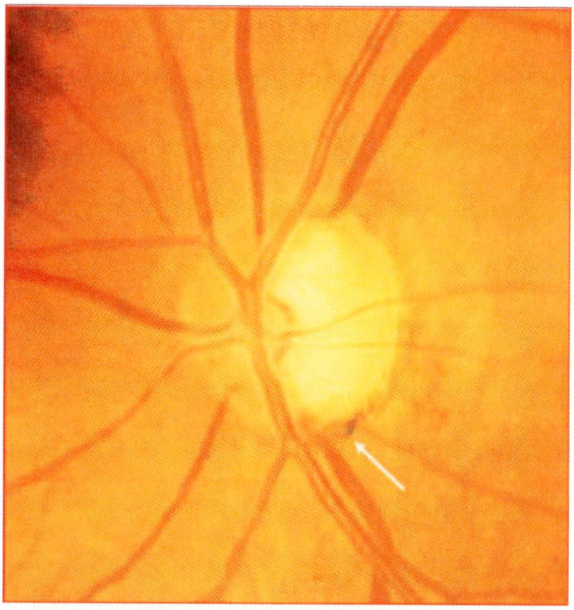

Fig. (14). Melanin (white arrow) was present in nearly six thousand patients studied in the observational study of the three leading causes of blindness.

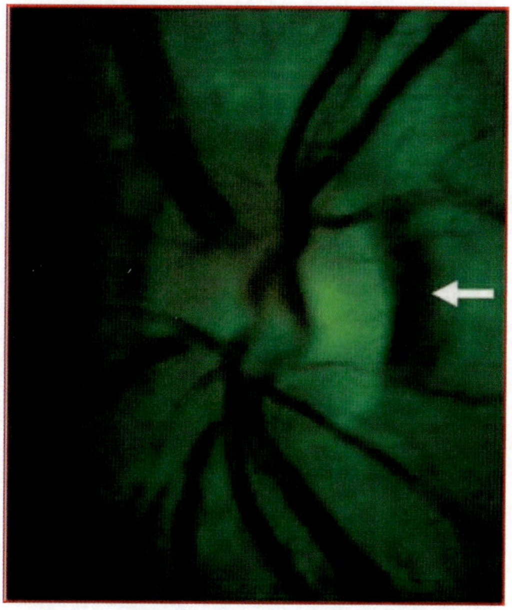

Fig. (15). With red-free light, melanin (white arrow) is observed easily.

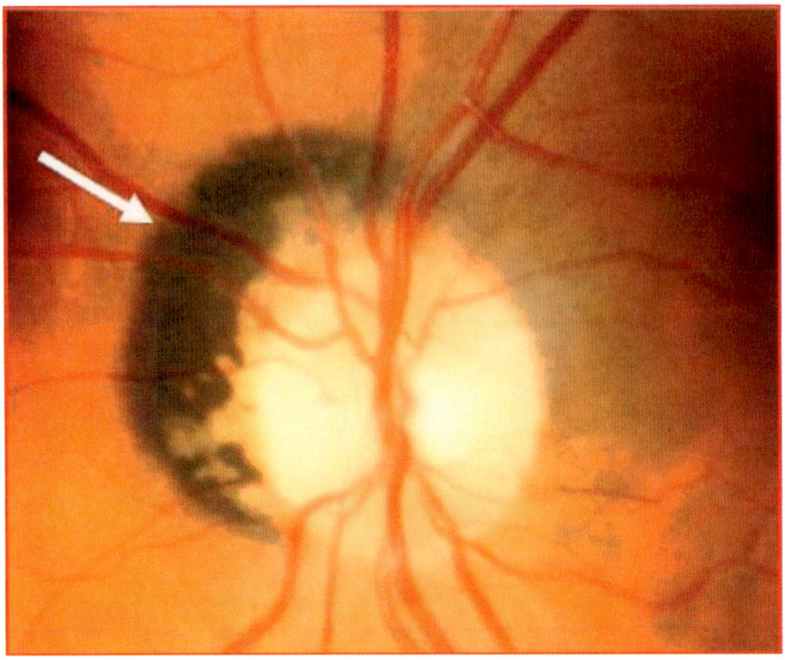

Fig. (16). In some patients, the presence of melanin (white arrow) is notorious.

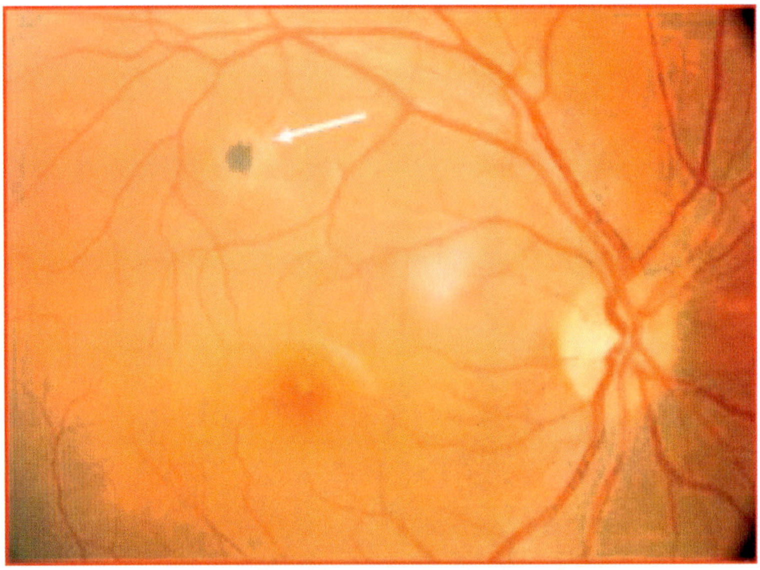

Fig. (17). Melanin (white arrow) can be observable in almost any area of the fundus.

Nature's insistence on placing melanin in virtually all patients examined gave rise to the questions: What nature wants sunscreen on a relatively obscure area? Why is it so important to nature?

A long-lasting search of the scientific literature on the physico-chemical properties and therefore the structure-activity relationship of melanin, did not resolve our doubts. None of the known properties of melanin successfully explained the nature insistence on placing melanin around the optic nerve in all patients, regardless of diagnosis. But in turn, later, we realized that none of the properties of melanin described in the literature could be considered as an obstacle, as something completely opposite, as something that will derail our finding about the intrinsic property of the melanin to dissociate and re-form the water molecule.

On the contrary, our findings concerning the ability of melanin to transform light energy into free chemical energy through the dissociation of the water molecule, is perfectly consistent with the usual unique properties of melanin which have been previously described. For two of the examples; it was a mystery how melanin absorbed dissipated energy, as they believed it was in the form of heat, as the vast majority of known substances, but gradually became apparent proving that it was not, for it was considered somehow dissipated melanin in a non-radiating energy absorbed, but did not understand how our finding about the intrinsic property of melanin to dissociate and re-form the water molecule fits perfectly;, the energy absorbed melanin dissipates through a non-radiative manner by means of the dissociation of the water molecule, which is an energy intensive process, as in the lab it requires a temperature of 2000 °C.

Another question resolved was why melanin's electronic behavior changes dramatically when hydrated. The answer is that the greater the amount of substrate present, greater dissociation or in other words as much light energy is converted freely into chemical energy.

Some observed details in the course of our study involving 6000 patients are important to explain which are as follows:

1. The omnipresence of melanin in each and every one of the six thousand patients studied at the time.
2. Despite being a tissue that requires 10 times more energy than the cerebral cortex, the photoreceptor layer, in normal conditions, does not contain a single blood vessel, which contradicts the dogma that all metabolic needs of the tissues are corrected by the stream blood, and therefore, hence require energy rods and cones? (Fig. **18**).
3. The fact that vascular endothelial cells seemed to respond to the greater or lesser amount and/or activity of melanin, because the larger the amount or activity of melanin, the lower number of blood vessels and conversely, the lower the amount or activity of melanin, the greater number of blood vessels,

but here we are talking about blood vessels whose morphological characteristics are not necessarily normal.

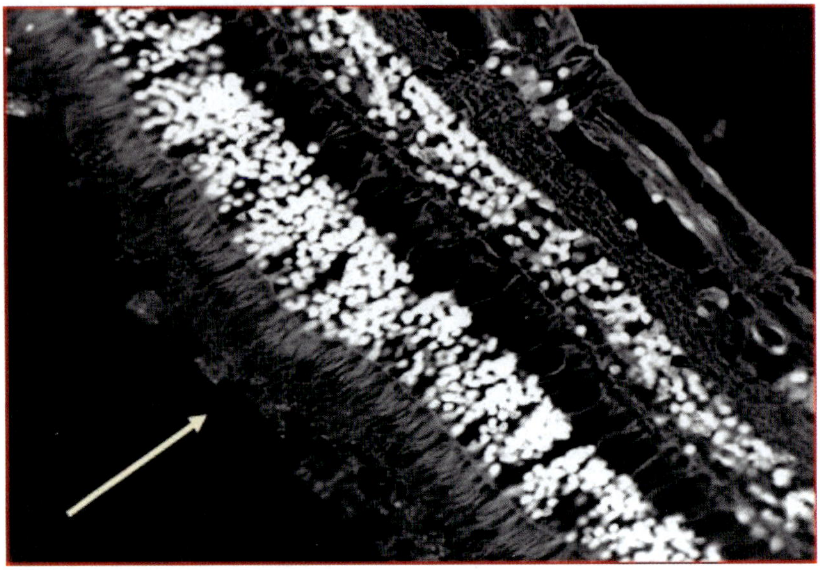

Fig. (18). The photoreceptor layer (arrow), seen in histological section of human retina; this layer requires 10 times more energy than cerebral cortex, six times than cardiac muscle, and three times than kidney cortex, and under normal conditions, no blood vessels at all. So from where energy has entered the photoreceptor layer?

A very interesting observation which has remained relatively unnoticed and unexplained until now, is that the difference in oxygen content of the blood retinal vein is 38% lower than in the retina arterial blood [60], but the difference between the venous and arterial blood of the choroid blood vessels is only 3%.

Therefore, it follows that the extraction of oxygen per milliliter of uveal blood is very low, a concept entirely theoretical, arguing that the metabolic needs of the choroid are small; however, explanation contradicts the fact that almost 80% of the oxygen consumed by the retina comes from the choroid [61].

Another argument that has been driven to try to explain the marked difference between the arterial and venous oxygen in both the retina and the choroid, is the high rate of blood flow in the choroid, which is approximately 27 times faster on it than on the retina [62]. It is said that the high rate of blood flow through the uvea produces a high pressure of oxygen in it, about 700 mmHg, thereby increasing its circulation to the retina [63].

The enormous speed of blood flow in the choroid does not seem to draw the attention of scholars; perhaps assume that the heart and blood vessels morphology are appropriate explanation. We think not, because we have found that the dissociation and re-formation of the water molecule produce a kind of vacuum around the vessels, which attract more blood flow (Fig. **19-A** and **19-B**).

But for researchers, the fact that the concentration of melanin in the choroid is 40% higher than in the skin, so much so that choroid means grape color, seems to go unnoticed, because that attitude is the usual melanin, since the presence of a single sunscreen is of no importance; and it is a phrase that is often heard: "*melanin does not deserve more attention*".

But if for a moment we try to explain the disparity between the oxygen content of venous blood and choroid and the enormous blood flow through the choroid taking into account the intrinsic property of melanin to dissociate and re-form the molecule water, a fact previously unknown, even unthinkable things would be more consistent, more coherent, for *e.g.* choroid having an amount several times more of melanin than the retina, it is expected that molecular oxygen content is several times more as it comes from the dissociation of the water molecule and not the lungs. Also, the molecular hydrogen content of the choroid and nearby tissues such as the photoreceptor layer must be several times greater than the tissues containing less melanin, but measurement thereof is a formidable challenge because hydrogen is the smallest atom (7 nanometers), so it is more practical at this time, the detection of their presence through their biological effects.

Fig. (19). 19-A and 19-B: The vacuum caused by the intrinsic property of melanin to dissociate and re-form the water molecule can be illustrated by this simple experiment in which they placed two similar containers with purified water, but left the container shown in the bottom fragments of melanin contained in silica to prevent combination with water. The deformation of the plastic packaging (PET) was noted to begin in 5 days; the photograph was taken six weeks after the experiment began.

The molecular oxygen we can measure in the blood or tissues with relative ease using electrodes sensitive to oxygen, but hydrogen is another story, for being diatomic (Fig. **20**), cannot be measured properly, for example, with the pH scale, as this measures the hydrogen ion, but not the diatomic, although discrete changes can be detected, for example from 7.4 to 7.1 in the pH scale by illuminating a sample of aqueous solution of melanin.

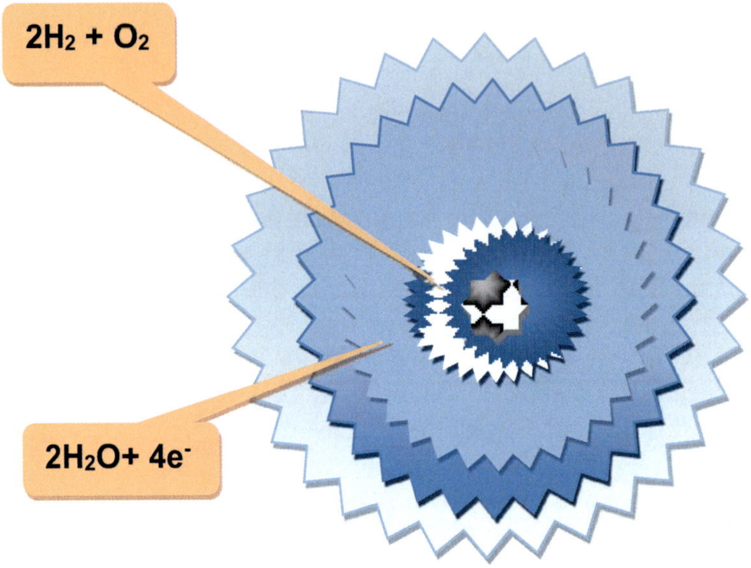

Fig. (20). The growing fields of energy emanating from the melanin molecule in the center of the diagram, whose nature is alternating, since a hydrogen containing diatomic and diatomic oxygen, and the next (and previous) containing water and 4 high-energy electrons per two water molecules re-formed. The shape and characteristic of these areas are fractal.

The regulation of rate of cell metabolism, according to Kleiber's Law; seems to be dictated by the same mechanism in all living beings, and intrinsic property of melanin to dissociate and re-form the water molecule, is the ideal candidate to be considered as the universal regulatory mechanism of the rate of cell metabolism so long sought, while is the cellular mechanism that explains the evolution amid entropy, which by law (the second), is rising inexorably.

The chemical reaction that we identified in the melanin is:

$$2H_2O \leftrightarrow 2H_2 + O_2 + 4e^-$$

Being the manner in which the melanin transforms visible and invisible light energy into chemical energy freely, and which is then carried by the diatomic hydrogen. The reaction occurs in the range of nano and picoseconds.

And in both phases (dissociation and re-formed) free chemical energy is released, which could be used in many ways by the cell. For example, the best known antioxidant is diatomic hydrogen, which continuously releases melanin, so that the interior of the cell is maintained with minimal oxidation.

The intracellular localization of melanin helps us to understand how regulation exerts metabolic rate of the cell; it is strategically placed in the perinuclear space, surrounding the cell nucleus [64] (Fig. **21**).

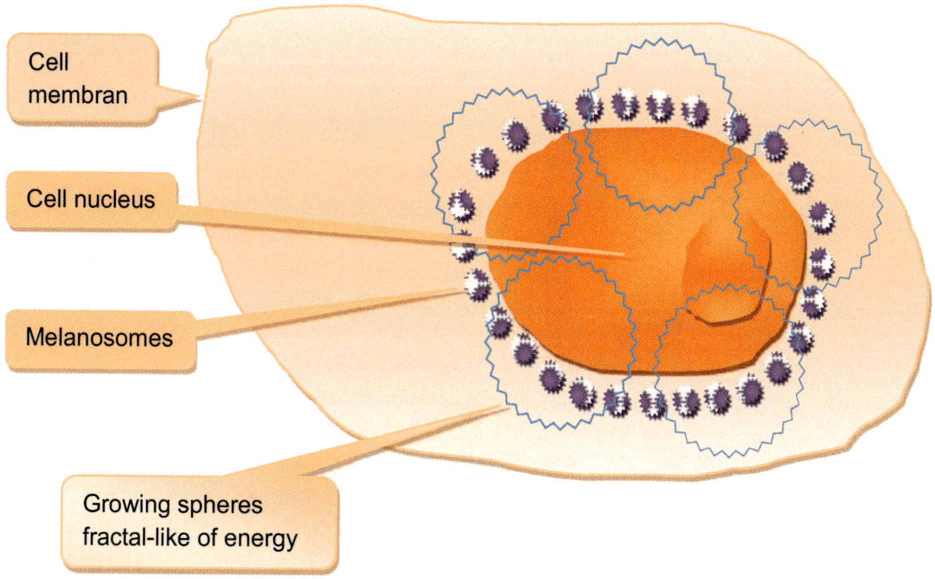

Fig. (21). The growing fields of energy melanin released, of fractal nature, that growing constantly, flooding the entire cell cytoplasm and its contents, *i.e.* all its organelles, and in areas where these fields converge for example the cell nucleus, forming a zone of high energy. Increasing energy spheres overlap each other, although this is not represented in the diagram.

Photo-system formed by light- melanin –water in the order of abundance in the universe, can be found in all living beings on the Earth, so we can infer that their function is the same in each and every one of them: the production of free chemical energy through dissociation of a water molecule, using force or energy that contains visible and invisible light.

The rate or frequency of this free chemical energy emanating from the melanin is relatively constant dictated by events, such as: the nature of light, the physicochemical characteristics of melanin, and the physicochemical characteristics of the water.

An idea of the stability of melanin is the fact that it has been identified in squid ink sacks fossilized and died 180 million years ago and is in good condition [65].

It can be said that a phenomenon that is as constant in their physicochemical characteristics, and to be universally present in all living beings on planet earth, adequately explain the constancy of a law turn constant of Kleiber's law or $\mathbf{E = M^{3/4}}$.

Beginning with the regulation of the rate of cell metabolism requires energy, and specifically free chemical energy available,the melanin is undoubtedly the perfect source. Do not forget that the universe is a delicate balance between energy and mass, and our body as well.

Therefore, in the case of preventive medicine, the potential impact of the balance of the initial reaction of life, *i.e.* the dissociation and re-formation of the water molecule and therefore dependent processes (all), is huge, more than you imagine.

Photosynthesis

The word photosynthesis means building or implementing something and then using force or energy that the light contains for it.

Fig. (22). The chlorophyll of plants, in the presence of light purple and red irreversibly dissociates in water molecule producing hydrogen and oxygen, both in molecular state.

The best known example occurs in the leaves of plants (Fig. **22**) where the chlorophyll in the presence of the ends of the visible light (purple and red) and water, irreversibly dissociated water molecule, generating hydrogen and oxygen diatomic *i.e.* chlorophyll is capable, at room temperature, to convert liquid water into gas, a process that requires heating in the laboratory water to two thousand degrees Celsius. Since hydrogen and oxygen are diatomic gases at room temperature.

Photosynthesis in Plants

The first reaction in the life of the plant is the dissociation of the water molecule, mainly producing hydrogen, which is the quintessential carrier of energy in the entire universe.

Chemical reaction is considered the most important in the world as it is the beginning of the food chain and can be represented as follows:

$$2H_2O \rightarrow 2H_2 + O_2$$

Actually chlorophyll is transforming photonic or light energy into free chemical energy. That is why the plant without water has not the substrate and thereby cannot transform light energy into free chemical energy and fails. The energy that is released to dissociate the water molecule is carried by the hydrogen, and this energy is what drives each and every one of the reactions that lead to a whole life. Photosynthesis in plants means the combination of carbon dioxide with water with glucose as the main product, as being the very first reaction of the dissociation of the water molecule.

Photosynthesis in Humans

It is a deeply rooted belief that only the chlorophyll of plants is able to dissociate the water molecule. But one of the findings of our observational study on the three leading causes of blindness in the world, which began in 1990 and ended in 2012, revealed the amazing ability of melanin, previously unknown, to dissociate and re-form the molecule of water.

The reaction can be represented as follows [66]:

$$2H_2O \leftrightarrow 2H_2 + O_2 + 4e^-$$

Water is oxidized in protons and also reducing factors (H_2 mainly) that are released, which in turn reduce the same (protons) forming water.

This means that our photosynthesis is more efficient than in the leaves of plants, because in them, it only dissociates irreversibly but our photo-system dissociates water and is also able to re-form it, something unique in nature. For each of the two molecules of re-formed water, there shall be four high-energy electrons.

Besides that, our chlorophyll, which is melanin, is able to absorb the entire electromagnetic spectrum, the visible and invisible light, from radio waves to gamma rays. We must not forget that melanin is the darkest substance known, because it absorbs all wavelengths without any reflection or refraction.

Melanin is also the most stable substance known, as it has been identified in squid ink sacks fossilized 180 million years old and in good condition [67]. The high concentration of melanin in the choroid layer (40% more than in the skin) became the answer to the mystery of the lack of vessels in the photoreceptor layer of the retina, perhaps because despite being more demanding, the human organism does not possess a single vessel under normal conditions.

Incidentally, choroid name means grape color due to high content of melanin (Fig. **23**).

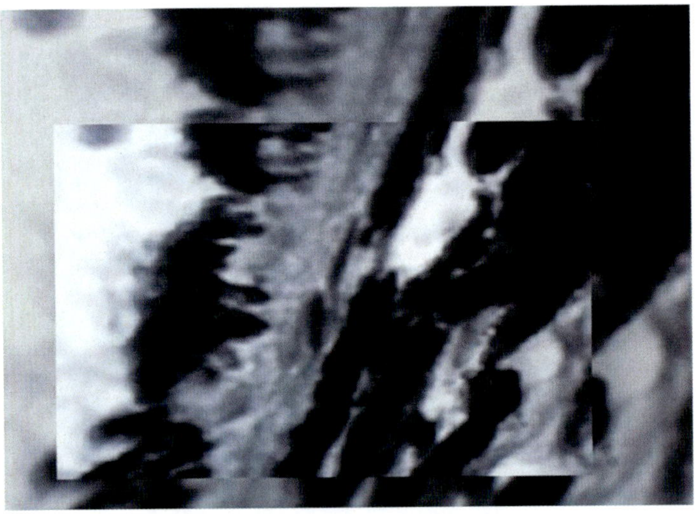

Fig. (23). Histological section human choroidal layer, everything dark is melanin, which is intracellular and the choroidal blood vessels are completely surrounded by cells with a high content of melanin; the blood vessels of choroidal layer have also a density per mm^2 of the most elevation in the body. To the right of the figure, cells can be seen -whose shape resembling certain arachnids- comprising the retinal pigment epithelium. Its basement membrane is called Bruch's membrane.

The answer to our question about where is the energy required photoreceptors was answered by the melanin in the choroid, as intrinsic property of dissociating and re-forming the water molecule, thereby producing free chemical energy explains how consistency and coherency with the evolution, functioning not only of the human eye but each and every one of the 100 trillion cells that make up tissues, organs and systems that the human body possesses.

And the free chemical energy levels of the tissues, which are determined by the activity of melanin, also explain our observation that the tissues respond with manners diverse to the presence of melanin. Since the retinal tissues of the living patient are observed with different characteristics depending on the amount and activity of melanin molecules present.

Glucose is a biomass source only. Since decades or centuries ago, it is a common finding in practically all textbooks that glucose acts both as a source of energy and at the same time as a source of starting material or building block for nearly all types of biosynthetic reactions [68]. Therefore, it is an idea that is deeply rooted, but on second thought, metaphorically speaking, if the energy source was glucose, then diabetic patients would have flown.

Glucose is present in all living things from the beginning of the times; so the insistence of nature about glucose function as metabolic intermediary or as a source of carbon chains must be critical, upmost important, but the concept as source of energy goes way beyond the physicochemical properties of the molecule.

Theoretically, glycolysis is a major energy production pathway used at least to some degree in all cells. Furthermore, glycolytic intermediates and products act as carbon sources for nearly all biosynthetic reactions, and the reducing equivalents required for most biosynthetic reactions are derived, at least in textbooks; from the flow of glucose through the pentose phosphate pathway. But for each and every one of the intermediate steps required for glycolysis, free chemical energy available is required previously to overcoming the energy of activation.

Glucose is out of discussion as the universal precursor of biomass; therefore, with glucose, our body synthesizes skin, nails, hair, muscle, bone, even to nucleic acids. But all this biomass only- not energy. The metabolism of glucose allows the cell and the organism access to chain carbon atoms, which are the backbone of most bio-molecules.

Although glucose can be phosphorylated in all tissues, which prevents the release of the cell, the reversal enzyme glucose-6-phosphatase is only found in liver and kidney. To phosphorylation and de-phosphorylation processes, free chemical energy *is sine qua non* required otherwise the necessary activation energy could not be overcoming for either of the two reactions occuring.

Gluconeogenesis is the reverse of glycolysis. Many of the enzymes involved are easily reversible, and are common to both pathways. Although reverse, each and every one of the steps required for gluconeogenesis, requires chemical free energy, and it is possible that depending on the degree of availability, the mechanism acts in a way or another. Interestingly, gluconeogenesis is inhibited by ethanol, which has diverse and widespread effects on the body, and worst, in any system, so when the damage or harmful effects are widespread; the first thing to think about is energy.

Finally, in regards Mitchell's chemiosmotic theory, the current explanation for the production of ATP by mitochondria; has not been resolved to date and is still just a theory.

ATP is Not the Perfect Currency for Energy Cell

The need for energy in the cell, to carry out its functions is huge. It is estimated that the human body requires 50-180 kg daily of ATP [69], as energy is required even to retain the shape up. But that average 50 g of ATP that the human body contains cannot explain ATP amplification, besides there is no way to do it according to the laws of conservation of energy. Moreover, each ATP molecule must be reconstituted at least 3 times per minute and it only requires energy.

And to complicate things, there is a classification of enzymes that not require ATP to perform its function, to which is added suffix synthase. But that does not mean they do not require free chemical energy to solve the energy of activation, simply means, using another form of energy that cannot be transported through the ATP.

Energy Cost of Protein Synthesis

The total cost can be estimated as the energy of four high-energy phosphate bonds for each peptide bond formed, *i.e.* per each amino acid polymerized. This does not include additional energy costs involved, *i.e.*, in DNA transcription. Since the free energy released in the hydrolysis of ATP into ADP amounts to approximately 30.5 kj/mol, and the hydrolysis of GTP is highly substrate dependent but comparable; we can conclude that the work needs to be performed by the cell in the process of adding an amino acid to a peptide sequence in the faithful performance of protein synthesis as at least 120 kj/mol, which exceeds by almost two orders of magnitude of the corresponding entropy reduction contribution. Other aspects that can be included in the total entropy analysis involve the change in the translational entropy of water that surrounds both DNA and protein surfaces and, as a result, loses several degrees of freedom per molecule. It has been demonstrated that a folding process such as protein folding, leads to a large amount of entropy increase [70].

Interestingly, it is becoming increasingly clear in cancer where molecular changes can be quantified in terms of entropy gain or information loss [71]. It is true that cell obeys the first law of thermodynamics, and simultaneously it locally acts against the second law of thermodynamics by creating structural and functional order. But mistakenly it is thought that cell creates and maintains information by expending energy produced from nutrient in the form of ATP and GTP, for the cell, nutrients, to get chains of carbon and other elements necessary to build biomass, but the energy obtained through the water, like plants.

Human Photosynthesis and Its Potential Implementation in PPPM

The ability to dissociate the water molecule starts to decrease from 26 years ahead in humans, in approximately 10% per decade, and after the 50s enters into free fall. Also decreases at any age with the cold, with contaminated water, iron supplements, with alcohol, with anesthetic agents, drugs with a high apparent volume of distribution, with pesticides, herbicides, with antidepressants, *etc.*

As well as in plants, in which all structures (stems, roots, leaves) rely on photosynthesis, ours is the same. All our processes are dependent on chemical energy that comes from photosynthesis. Despite that our body, after 4 billion years of evolution, knows its job very well, it cannot do that when photosynthesis level is not adequate; being the major reason by which diseases develop, in other words: when the free chemical energy that emanates from the melanin is not sufficient or adequate then body goes under disease.

Some Examples of Potential Applications of the Human Photosynthesis Enhancement

In our laboratory, we have identified compounds which possess the intrinsic ability to intensify the process we call human photosynthesis for his analogy with plants. And once the scientific, ethical and legal requirements regarding it were filled, then we began to use in treatment in one of the leading causes of blindness: macular degeneration related to age (AMD) (Fig. **24**).

Age-related Macular Degeneration (AMD): Alzheimer´s Disease in the Eye [72]?

Age-related macular degeneration (AMD) is the leading cause of visual impairment in elderly persons in industrialized countries. Alzheimer´s disease and AMD share a common pathogenesis based on several lines of evidence [73]. Both conditions have similar histopathological changes. In early AMD, an accumulation of drusen containing extracellular β-amyloid, lipid, and other waste products derived from degenerating neuroretina have been documented. In Alzheimer´s disease, an accumulation of extracellular β-amyloid, axonal, and dendritic waste products from dystrophic neurons have been also documented.

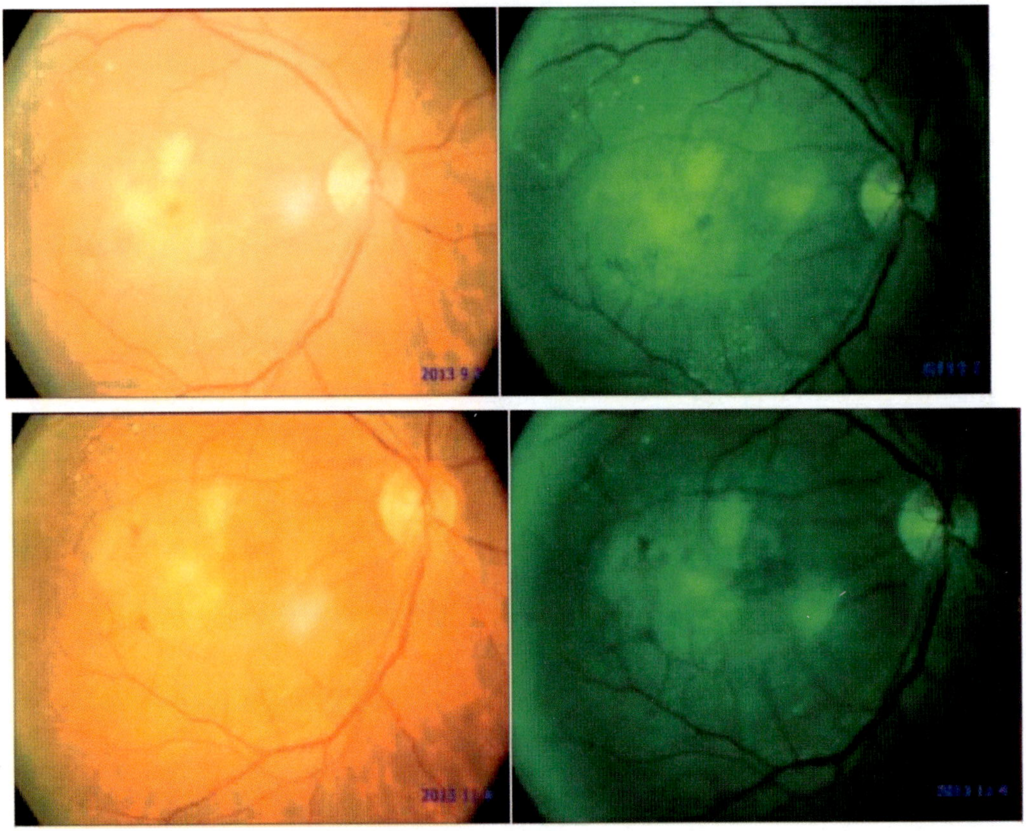

Fig. (24). This is female patient 78 years old with vision loss in OD of a year of evolution. Human photosynthesis orally intensified by the compounds developed in our laboratory. At 8 weeks of treatment of macular edema markedly decreased by allowing the anatomy of the fovea, the central part of the macula, is significantly recovering. Images from the top correspond to the examination on September 2, 2013; photographs of the bottom were taken on November 4, 2013. In the red-free photography, the improvement is evident in the edema of macula.

These deposits form senile plaques and neurofibrillary tangles in the cortex and hippocampus of the brain that lead to neuronal malfunction and cell death in the later stages of Alzheimer´s disease. Clinical studies suggest that AMD and AD share similar vascular risk factors, such as hypertension and cigarette smoking.

In the aging population, two of the most common neurodegenerative diseases, AD and AMD, could be simultaneously treated or prevented with a single intervention

[74]. And our therapeutic results in both diseases have been very successful in both cases, both AD and AMD (Figs. **25** and **26**).

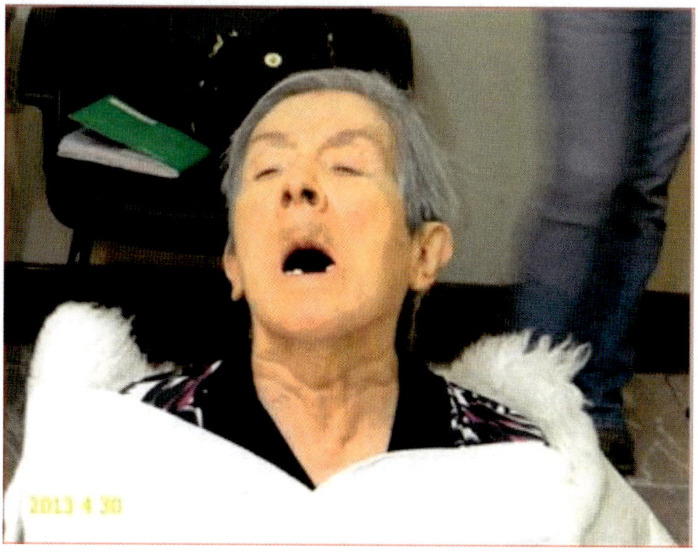

Fig. (25). Female patient 83 years old with AD 7 years of evolution, and not talking, not walking, using diaper, and do not open her eyes. It was first seen on April 30, 2013. That day home treatment was started, orally, at the rate of three drops under the tongue every two hours during the day.

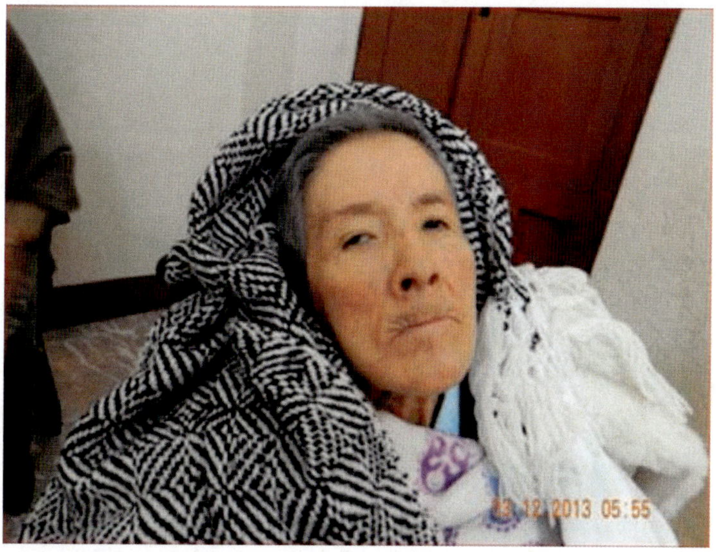

Fig. (26). The picture is from the same patient as in (Fig. **25**), when it was examined the day of December 3, 2013, after 8 months of treatment. The improvement is remarkable, because the patient now pays attention, and says some things, answers to simple questions, no longer has an open mouth all the time. The constant lung and bowel problems disappeared.

Our finding of the previously unknown capacity of the melanin´s human body to split and re-form the water molecule, appointed as human photosynthesis by its analogies with photosynthesis in plants, constitutes a disruptive advance in the knowledge because it means that glucose is not a source of energy, instead it is a biomass source only. If really glucose was a source of energy, then diabetic patients would have been able to fly. It is out of discussion that glucose is the ideal building block; our body even makes nucleic acids with it. But energy (everything that produces a change) is astonishingly taken by the cell from water, as happens in plants.

AD is an energy problem, not a biomass trouble [75].

Neurotransmitters Role

Expression of the appropriate neurotransmitters is essential for the function of neural circuits; however sufficient free chemical energy is fundamental at most. Neurons are able to change their transmitter phenotype as result and to deal with in the best way with variations mainly of the light levels in the environment. Variables other than humidity, temperature, pH, food availability (biomass source), age, sex, stress degree and so forth are important too; but the amount of light, visible and invisible, are important especially due to the relatively recent finding of the intrinsic property of melanin to split and re-form the water molecule [76], thereby melanin possesses the unexpected capacity to transform the photonic energy into free chemical energy that is susceptible to be used by the eukaryotic cell.

Neurotransmitter expression switched in hypothalamic neurons is not an isolated case. Transmitter switching occurs at several levels: at transcriptional level and is accompanied by changes in postsynaptic receptors. The very first common requirement at all levels is the sufficient amount of free chemical energy available, otherwise, the highly complex neuronal cell organization cannot carry out a proper neurotransmitter expression switching, owing to every single part of the large number of processes involved [80].

CONCLUSION

Photosynthesis is basically a black box, the internal mechanism of which is poorly understood. Photo-transduction or the transformation of the photonic energy (coming from light) into free chemical energy is a very difficult process to study, because, like other metabolic processes, photosynthesis proved highly sensitive to intervention [77]. Whenever the covering of the box is pried off, the wheels inside stop turning [78]. Outside the living cells, all the molecules that take part in the photosynthetic processes must be rendered reactive, because they are known to be highly inert.

Photosynthesis reactions are surface-dependent in plants as well in humans. It is possible that light absorption itself is not sensitive to anesthetics; instead, damage must involve some kind of surface-sensitive reaction. Recalling the fact that normal aged patients sometimes awake from a general anesthesia procedure with Alzheimer´s disease.

Challenges of Current Healthcare Systems

Currently, severe chronic pathologies such as diabetes mellitus, neurologic degenerations, cardiovascular disorders and cancer are treated after the onset of the disease. Pessimistic or perhaps realistic prognosis considers pandemic scenario for type 2 DM, neurodegenerative disorders, cardiovascular disorders and some types of cancer over the next 10-20 years accompanied by severe economic burden for the European healthcare systems and worldwide.

Advanced healthcare aims to promote a paradigm shift from delayed interventional to predictive medicine tailored to the person, from reactive to preventive medicine and from disease wellness. The cost-effective management of diseases and the crucial role of predictive, preventive and personalized medicine (PPPM) in the modernization of healthcare have been acknowledged as priorities by global and regional health-related institutions.

The integrative concept raised by EPMA [79] would enable clinicians to predict and provide targeted preventive measures before the actual onset of the disease. The expected outcomes are conducive to more effective population screening, prevention measures early in life, identification of populations and persons at-risk, stratification for optimal therapy planning, prediction and reduction of adverse drug-drug, drug-disease, and drug-physiology interactions [80].

The use of emerging technologies will be more effective if we redirect the current concepts of cell biology and begin to stop thinking that water is only something like an inner cleaner, it's not just a liquid that comes clean and comes out dirty, that our body has used. Plasmatic volume, organic base fluids (saliva, plasma, intracellular environment, *etc.*) instead are an essential part of the crucial process of obtaining energy.

The algorithms that can be planted in order to improve the health of the population, will be really efficient when we stop to think that glucose is a source of energy.

Being aware that the sacred role of glucose as an energy source of the body now breaks into a thousand pieces and they really are only biomass source, we will be more aware of the importance of the role of water in cell physiology as electrons source.

For health care and disease prevention, it is essential that the intrinsic property of melanin to dissociate and re-form the molecule of water remains in the right stable state.

If we assume that the human body is the result of nearly four billion years of evolution, we can infer that our body knows its job perfectly, as it has done since millions of years.

Given, this should surprise the sick; apparently our bodies make mistakes when there are failures at the molecular level and are amplified until signs or symptoms of disease appear. The cell requires an adequate supply of energy, *i.e.*, constant, continuous, within the same range that is over four billion years of evolution. Otherwise, the cell dies in minutes, as the continuous decline in entropy. For essential life and health process energy expenditure is required that the theory of ATP and GDP cannot explain, but the intrinsic property of melanin to dissociate and re-form the molecule of water, thus transforming the light or photon energy in free chemical energy. It's a kind of process that can explain the energetic requirements that cell needs to decrease the entropy and make life (and health) possible.

If we assume that the entropy decreases the tendency to disorder, thereby life and health are possible to maintain, and then we can think also the opposite: when the entropy decreases, health is threatened and diseases appear.

In other words, when energy levels are adequate, the organization is perpetuated, and when energy levels are altered, the complex cell function will also be altered, which is not compatible with our concept of health.

We believe that the discovery of the intrinsic chemistry of melanin marks a turning point, as the energy emanating from melanin, being the fundamental role of the cell, can be said to control the entire cell function. One of the important aspects of the discovery of human photosynthesis, is the fact that a single reaction interacts with virtually all others, as well as in the tree, where from root to the farthest leaf stem, and are governed depending on photosynthesis, and which is similar in our body as well.

As a new horizon of possibilities now opens since the pharmacological modulation of a single chemical reaction: the dissociation and re-built of water molecule that can influence decisively in a good number of known diseases to date. There is no doubt that curative and preventive medicine will no longer be the same after the discovery of human photosynthesis.

And according to the current trend of making preventive medicine really effective, the real possibility of pharmacologically modulating the basis of the life process, constitutes a powerful tool that will allow us to stay healthy, for the steady state of the dissociation and re-formation of the molecule of water, enabling adequate levels of chemical free energy to cell.

The disclosure of the unsuspected physiological role of melanin for both the public and the medical professionals is a challenge to overcome, as it is a conceptual revolution of biblical proportions, but it is an essential step on the long road to implement as massive pharmacological modulation of human photosynthesis in the exercise of both preventive and curative medicine, what undoubtedly marks a before and after.

Understanding of a process strictly regulated by 4 billion years of evolution is required to observe the phenomenon both for patients and health professionals so that information about critical aspects is generated and disseminated, in order to clarify and eventually allow and validate the concept through large-scale population-based future studies, and whose preliminary names could be:Human Photosynthesis and Neurodegeneration or Human photosynthesis and Diabetes Mellitus.

Outlook

The intrinsic property of melanin to split and reform the water molecule, termed as human photosynthesis by analogy with plants, is a model capable of accounting for all functions of eukaryotic cell. Therefore, its therapeutic possibilities are enormous, both from the point of view of reactive or curative medicine and of preventive, predictive and personalized medicine [81].

The potential implementation of pharmacological modulation of human photosynthesis is a relatively simple process compared with the current trend, as it does not require large investments in facilities and equipment diagnostics, and initial unnecessary stress because the physician and patient search for a diagnosis, takes second place, because in reality to the human body the names that we put to diseases have absolutely no relevance.

Through medical intensification of the process of dissociation and re-formation from the water, the body gets an adequate energy balance and begins to function as it has since millions of years.

HUMAN AND ANIMAL RIGHTS

No humans were used in this research. All animal research procedures followed were in accordance with the standards set forth in the eighth edition of Guide for the Care and Use of Laboratory Animals published by the National Academy of Sciences, The National Academies Press, Washington, D.C.

CONFLICT OF INTEREST

The author (editor) declares no conflict of interest, financial or otherwise.

ACKNOWLEDGEMENTS

Dr. Arturo Solís Herrera received the support of the Center for studies of Photosynthesis in Humans®, S.C.

REFERENCES

[1] Portet, S.; Tuszynski, J.A.; Dixon, J.M.; Sataric, M.V. Models of spatial and orientational self-organization of microtubules under the influence of gravitational fields. *Phys. Rev. E Stat. Nonlin. Soft Matter Phys.*, **2003**, *68*(2 Pt 1), 021903. [http://dx.doi.org/10.1103/PhysRevE.68.021903] [PMID: 14525002]

[2] Gordon, R.; Goel, N.S.; Steinberg, M.S.; Wiseman, L.L. A rheological mechanism sufficient to explain the kinetics of cell sorting. *J. Theor. Biol.*, **1972**, *37*(1), 43-73. [http://dx.doi.org/10.1016/0022-5193(72)90114-2] [PMID: 4652421]

[3] Van den Bosch, F. http://www.astro.yale.edu/vdbosch/astro320.html

[4] Isaac, A.S. In the game of Energy and Thermodynamics you can′t break even. *Smith. Inst. J.*, 1970 June;, 6.

[5] Lindsay, R.B. Physics-to what extent is it deterministic? *Am. Sci.*, **1968**, *56*, 100.

[6] Julian, H. *Evolution and Genetics in What is Man*; Newman, J.R., Ed.; Simon and Schuster: New York, **1955**, p. 278.

[7] Morris, J.D. Can Scientists Study the Past? *Acts Facts,* **1992**, *21*(8) Available at: http://www.icr.org/article/can-scientists-study-past/

[8] Henry, M.; Duane, G. *The Battle for Creation.*, **1976**. http://www.icr.org/article/young-earth-creationist-bibliography/

[9] Portet, S.; Tuszynski, J.A.; Dixon, J.M.; Sataric, M.V. Models of spatial and orientational self-organization of microtubules under the influence of gravitational fields. *Phys. Rev. E Stat. Nonlin. Soft Matter Phys.*, **2003**, *68*(2 Pt 1), 021903. [http://dx.doi.org/10.1103/PhysRevE.68.021903] [PMID: 14525002]

[10] Gordon, R.; Goel, N.S.; Steinberg, M.S.; Wiseman, L.L. A rheological mechanism sufficient to explain the kinetics of cell sorting. *J. Theor. Biol.*, **1972**, *37*(1), 43-73. [http://dx.doi.org/10.1016/0022-5193(72)90114-2] [PMID: 4652421]

[11] Douglas, D. *Dictionary of Mathematics Terms,* 3rd ed; Barron's Educational Series: New York, **2009**. http://sygmabeaute.com/download-pdf/6988-dictionary-of-mathematics-terms-3rd-edition-barrons-professional-guides-series.html

[12] Davies, P.C.; Rieper, E.; Tuszynski, J.A. Self-organization and entropy reduction in a living cell. *Biosystems,* **2013**, *111*(1), 1-10.

[http://dx.doi.org/10.1016/j.biosystems.2012.10.005] [PMID: 23159919]

[13] Schrödinger, E. *What is Life? The physical aspects of the Living Cell.*, **1967**. http://www.worldcat.org/title/what-is-life-the-physical-aspect-of-the-living-cell-mind-and-matter/oclc/42984716

[14] Speksnijder, J.E.; Sardet, C.; Jaffe, L.F. Periodic calcium waves cross ascidian eggs after fertilization. *Dev. Biol.*, **1990**, *142*(1), 246-249. [http://dx.doi.org/10.1016/0012-1606(90)90168-I] [PMID: 2227098]

[15] Erie, J.C.; Butz, J.A.; Good, J.A.; Erie, E.A.; Burritt, M.F.; Cameron, J.D. Heavy metal concentrations in human eyes. *Am. J. Ophthalmol.*, **2005**, *139*(5), 888-893. [http://dx.doi.org/10.1016/j.ajo.2004.12.007] [PMID: 15860295]

[16] Dräger, U.C.; Balkema, G.W. Does melanin do more than protect from light? *Neurosci. Res. Suppl.*, **1987**, *6* (Suppl. 6), S75-S86. [http://dx.doi.org/10.1016/0921-8696(87)90009-0] [PMID: 3317148]

[17] Jimbow, K. Current update and trends in melanin pigmentation and melanin biology. *Keio J. Med.*, **1995**, *44*(1), 9-18. [http://dx.doi.org/10.2302/kjm.44.9] [PMID: 7760535]

[18] Tosti, A.; Costelli, E. The function of melanocytes. In: *Melanoma, Biology and Management*; Cascinelli, N.; Santinami, M.; Veronesi, U., Eds.; Masson: Milan, **1990**; pp. 75-77.

[19] Boulton, M. Melanin and the retinal pigment epithelium In: *The Retinal Pigment Epithelium*; Michael, E.; Marmor Thomas, J.; Wolfenberger, , Eds.; Oxford University Press: New York, Oxford, **1998**.

[21] Arturo, S.H.; María del Carmen, A.E.; Martha Patricia, S.A. *Photosynthesis in Humans. Chapter 12 of the book Photosynthesis, Functional Genomics, Physiological Processes, and Environmental Issues*; Nafees, Khan, Ed.; Nova Publishers: New York, **2015**.

[22] Meredith, P.; Sarna, T. The physical and chemical properties of eumelanin. *Pigment Cell Res.*, **2006**, *19*(6), 572-594. [http://dx.doi.org/10.1111/j.1600-0749.2006.00345.x] [PMID: 17083485]

[23] Hill, H.Z. The function of melanin or six blind people examine an elephant. *BioEssays*, **1992**, *14*(1), 49-56. [http://dx.doi.org/10.1002/bies.950140111] [PMID: 1546980]

[25] Sealy, R.C. Free radicals in melanin formation, structure, and reactions. In: *Free radicals in molecular biology, aging, and disease*; Armstrong, D., Ed.; Raven: New York, **1989**; pp. 67-76.

[26] Crippa, P.R.; Cristofoletti, V.; Romeo, N. A band model for melanin deducted from optical absorption and photoconductivity experiments. *Biochim. Biophys. Acta*, **1978**, *538*(1), 164-170. [http://dx.doi.org/10.1016/0304-4165(78)90260-X] [PMID: 563737]

[27] Korytowski, W.; Pilas, B.; Sarna, T.; Kalyanaraman, B. Photoinduced generation of hydrogen peroxide and hydroxyl radicals in melanins. *Photochem. Photobiol.*, **1987**, *45*(2), 185-190. [http://dx.doi.org/10.1111/j.1751-1097.1987.tb05362.x] [PMID: 3031709]

[28] Post, P.W.; Daniels, F., Jr; Binford, R.T., Jr Cold injury and the evolution of "white" skin. *Hum. Biol.*, **1975**, *47*(1), 65-80. [PMID: 1126703]

[29] Menon, I.A.; Persad, S.; Ranadive, N.S.; Haberman, H.F. Role of superoxide and hydrogen peroxide in cell lysis during irradiation *in vitro* of Ehrlich ascitic carcinoma cells in the presence of melanin. *Can. J. Biochem. Cell Biol.*, **1985**, *63*(4), 278-283. [http://dx.doi.org/10.1139/o85-041] [PMID: 2990646]

[30] Matsuoka, L.Y.; Wortsman, J.; Haddad, J.G.; Kolm, P.; Hollis, B.W. Racial pigmentation and the cutaneous synthesis of vitamin D. *Arch. Dermatol.*, **1991**, *127*(4), 536-538. [http://dx.doi.org/10.1001/archderm.1991.04510010104011] [PMID: 1848745]

[31] Mizutani, U.; Massalski, T.B.; McGinness, J.E.; Corry, P.M. Low temperature specific heat anomalies

in melanins and tumour melanosomes. *Nature,* **1976**, *259*(5543), 505-507. [http://dx.doi.org/10.1038/259505a0] [PMID: 1256550]

[32] Osak, W.; Tkacz, K.; Slawinski, J.; Czternastek, H. Dielectric relaxation in synthetic melanin. *Biopolymers,* **1989**, *28*, 1875-1833. [http://dx.doi.org/10.1002/bip.360281104]

[33] Steel, K.P.; Barkway, C. Another role for melanocytes: their importance for normal stria vascular is development in the mammalian inner ear. *Development,* **1989**, *107*(3), 453-463. [PMID: 2612372]

[34] Barrenäs, M.L.; Lindgren, F. The influence of inner ear melanin on susceptibility to TTS in humans. *Scand. Audiol.,* **1990**, *19*(2), 97-102. [http://dx.doi.org/10.3109/01050399009070759] [PMID: 2371541]

[35] Crippa, P.R.; Viappiani, C. Photoacoustic studies of non-radiative relaxation of excited states in melanin. *Eur. Biophys. J.,* **1990**, *17*(6), 299-305. [http://dx.doi.org/10.1007/BF00258378] [PMID: 2307137]

[36] Crippa, P.R.; Mazzini, A.; Salmelli, D. Oxidation of NADH by melanin: effect of UV light and copper ions. *Physiol. Chem. Phys.,* **1979**, *11*(6), 491-499. [PMID: 232923]

[37] Jimbow, K. Current update and trends in melanin pigmentation and melanin biology. *Keio J. Med.,* **1995**, *44*(1), 9-18. [http://dx.doi.org/10.2302/kjm.44.9] [PMID: 7760535]

[38] Prota, G. *Melanins and Melanogenesis*; Academic Press: San Diego, London, **1992**. Elsevier Inc. 1992 ISBN: 978-0-12-565970-3

[39] Nordlund, J.J.; Abdel-Malek, Z.A.; Boissy, R.E.; Rheins, L.A. Pigment cell biology: an historical review. *J. Invest. Dermatol.,* **1989**, *92*(4) Suppl., 53S-60S. [http://dx.doi.org/10.1038/jid.1989.33] [PMID: 2649615]

[40] Childs, B. Sir Archivaid Garrod's conception of chemical individuality: A modern appreciation. *New Eng. J. Med.,* **1970**, *282*, 71-77.

[41] Witkop, C.J.J. *Albinism in Advances in Human Genetics*; Harris, H.; Hirschhorn, K., Eds.; Plenum Press: New York, **1971**, pp. 61-142.

[42] Abadi, R.; Pascal, E. The recognition and management of albinism. *Ophthalmic Physiol. Opt.,* **1989**, *9*(1), 3-15. [http://dx.doi.org/10.1111/j.1475-1313.1989.tb00797.x] [PMID: 2512527]

[43] Apkarian, P.; Eckhardt, P.G.; van Schooneveld, M.J. Detection of optic pathway misrouting in the human albino neonate. *Neuropediatrics,* **1991**, *22*(4), 211-215. [http://dx.doi.org/10.1055/s-2008-1071443] [PMID: 1775217]

[44] Conlee, J.W.; Bennett, M.L. Turn-specific differences in the endocochlear potential between albino and pigmented guinea pigs. *Hear. Res.,* **1993**, *65*(1-2), 141-150. [http://dx.doi.org/10.1016/0378-5955(93)90209-J] [PMID: 8458747]

[45] Silvers, K. Willys. *The coat color of Mice, a model of mammalian gene action and interaction*; , **1979**. [http://dx.doi.org/10.1007/978-1-4612-6164-3]

[46] Prota, G.; Alessandra, N. *The Chemistry of Melanins and Related Metabolites. Chapter 24 of the book: The Pigmentary System. Physiology and Pathophysiology*; Nordlund, J.J.; Boissy, R.E.; Hearing, V.J.; King, R.A.; Ortonne, J.P., Eds.; Oxford University Press: Oxford, New York, **1998**.

[47] Halaban, R.; Moellmann, G. Murine and human b locus pigmentation genes encode a glycoprotein (gp75) with catalase activity. *Proc. Natl. Acad. Sci. USA,* **1990**, *87*(12), 4809-4813. [http://dx.doi.org/10.1073/pnas.87.12.4809] [PMID: 1693779]

[47b] Carraro, C.; Pathak, M.A. Characterization of superoxide dismutase from mammalian skin epidermis.

J. Invest. Dermatol., **1988**, *90*(1), 31-36.
[http://dx.doi.org/10.1111/1523-1747.ep12462534] [PMID: 3335787]

[48] Schallreuter, K.U.; Wood, J.M. Thioredoxin reductase in control of the pigmentary system. In: *Clinics in Dermatology: Disorders of Pigmentation*; Mackie, R., Ed.; J. B. Lippincott: Philadelphia, **1989**; pp. 92-105.
[http://dx.doi.org/10.1016/0738-081X(89)90060-6]

[49] Sarna, T.; Swartz, H.M. *Chapter 25 of the book: The Pigmentary System*; Nordlund, J.J.; Boissy, R.E.; Hearing, V.J., Eds.; Oxford University Press: New York, Oxford, **1998**.

[50] Wolf, K. Melanocyte-keratinocyte interactions *in vivo*. The fate of melanosomes. *Yale J. Biol. Med*, **1973**, *46*, 384-396.

[51] Mann, D. M. A. Lipoprotein pigments –their relationships to aging in the human neuron system. II. The melanin content of pigmented nerve cells. *Brain*, **1974**, *97*, 489-498.

[52] Commoner, B.; Townsend, J.; Pake, G.E. Free radicals in biological materials. *Nature*, **1954**, *174*(4432), 689-691.
[http://dx.doi.org/10.1038/174689a0] [PMID: 13213980]

[53] Crippa, P.R. Chemistry of melanins In: *The Alkaloids*; Brossi, A.; Horak, V.; Prota, G.; Svoronos, P.; Wolfram, L., Eds.; Academic Press: New York, **1989**; pp. (36)253-323.

[54] Siegfried, C.J.; Shui, Y-B. Racial differences in ocular oxidative metabolism. *Arch. Ophthalmol.*, **2011**, *129*(7), 849-854.
[http://dx.doi.org/10.1001/archophthalmol.2011.169] [PMID: 21746975]

[55] Kleiber, M. Body size and metabolism. *Hildargia*, **1932**, *6*(11), 315-353.
[http://dx.doi.org/10.3733/hilg.v06n11p315]

[56] McMahon, T. Size and shape in biology. *Science*, **1973**, *179*(4079), 1201-1204.
[http://dx.doi.org/10.1126/science.179.4079.1201] [PMID: 4689015]

[57] Kleiber, M. *The Fire of Life*; Wiley: New York, **1961**.

[58] Geoffrey, W. *Santa Fe Institute, in the e-book: Fractals, hunting down the hidden dimension*; NOVA Publishers, **2011**.

[59] David, H. Chapter 3 of the e- book *Eye, Brain, and Vision*, **1995**. http://hubel.med.harvard.edu

[60] Hickam, J.B.; Frayser, R.; Ross, J.C. A study of retinal venous blood oxygen saturation in human subjects by photographic means. *Circulation*, **1963**, *27*, 375-385.
[http://dx.doi.org/10.1161/01.CIR.27.3.375] [PMID: 13961118]

[61] Alm, A.; Bill, A. The oxygen supply to the retina. II. Effects of high intraocular pressure and of increased arterial carbon dioxide tension on uveal and retinal blood flow in cats. A study with radioactively labelled microspheres including flow determinations in brain and some other tissues. *Acta Physiol. Scand.*, **1972**, *84*(3), 306-319. b
[http://dx.doi.org/10.1111/j.1748-1716.1972.tb05182.x] [PMID: 4553229]

[62] Alm, A.; Bill, A.; Young, F.A. The effects of pilocarpine and neostigmine on the blood flow through the anterior uvea in monkeys. A study with radioactively labelled microspheres. *Exp. Eye Res.*, **1973**, *15*(1), 31-36.
[http://dx.doi.org/10.1016/0014-4835(73)90186-3] [PMID: 4630582]

[63] Bill, A.; Sperber, G.; Ujiie, K. Physiology of the choroidal vascular bed. *Int. Ophthalmol.*, **1983**, *6*(2), 101-107.
[http://dx.doi.org/10.1007/BF00127638] [PMID: 6403480]

[64] Cotzias, GC; Procter, A *CRC Handbook on Free Radicals and Human Disease. A review*; CRC Handbook of Free Radicals and Antioxidants, **1989**, *vol. 1*, pp. 209-221.

[65] Goldsmith, L.A. *Biochemistry and Physiology of the Skin*; Volumes 1 and II. Oxford Medicine

Publications, Oxford University Press. **1983**. ISBN 0192612530

[66] Arturo , S.H. Photo-electrochemical method of separating water into hydrogen and oxygen, using melanins or the analogues, precursors or derivatives thereof as the central electrolyzing element. US 8,455,145 B2, **2013**.

[67] Fitzpatrick, T.B.; Szabó, Wick MM. Biochemistry and physiology of melanin pigmentation. In: *Biochemistry and physiology of the skin*; Goldsmith, L.A., Ed.; Oxford University Press: New York, **1983**; pp. 687-712.

[68] Stryer Chapter 30: *Integration of Metabolism*. In: *Biochemistry,* 4th; W.H. Freeman & Company: New York, **1995**.

[69] Jerry, B. ATP the perfect energy currency for the cell. *Creat. Res. Soc. Quarte.,* **1999**, *36*, 1.

[70] Ivan, R The structural basis of the myosin ATPase activity. *J. Biol. Chem.,* **1996**, *271*(27), 15850-15853.

[71] Kinoshita, M. Importance of translational entropy of water in biological self-assembly processes like protein folding. *Int. J. Mol. Sci.,* **2009**, *10*(3), 1064-1080. [http://dx.doi.org/10.3390/ijms10031064] [PMID: 19399238]

[72] Frieden, B.R.; Gatenby, R.A. Information dynamics in living systems: prokaryotes, eukaryotes, and cancer. *PLoS One,* **2011**, *6*(7), e22085. [http://dx.doi.org/10.1371/journal.pone.0022085] [PMID: 21818295]

[73] Kaarniranta, K.; Salminen, A.; Haapasalo, A.; Soininen, H.; Hiltunen, M. Age-related macular degeneration (AMD): Alzheimer's disease in the eye? *J. Alzheimers Dis.,* **2011**, *24*(4), 615-631. [PMID: 21297256]

[74] Baker, M.L.; Wang, J.J.; Sophie, R.; Ronald, K.; Lewis, H.K.; Larsen, A.K. ; Tien Yin, W. Early age-related macular degeneration, cognitive function, and dementia. *Arch. Ophthalmol.,* **2009**, *127*(5), 667-673. [http://dx.doi.org/10.1001/archophthalmol.2009.30] [PMID: 19433718]

[75] Ding, J.D.; Lin, J.; Mace, B.E.; Herrmann, R.; Sullivan, P.; Bowes, R.C. Targeting age-related macular degeneration with Alzheimer's disease based immunotherapies: anti-amyloid-β antibody attenuates pathologies in an age-related macular degeneration mouse model. *Vision Res.,* **2008**, *48*(3), 339-345. [http://dx.doi.org/10.1016/j.visres.2007.07.025] [PMID: 17888483]

[76] Solis-Herrera, A.; Leszek, J.; del Carmen Arias Esparza, M.; Solís-Arias, RI.; Solís-Arias, P.E. Human photosynthesis: A turning point in the understanding and treatment of alzheimer´s disease. *J. Bioanal. Biomed.,* **2013**, *5*, 057-060.

[77] Dulcis, D.; Jamshidi, P.; Leutgeb, S.; Spitzer, N.C. Neurotransmitter switching in the adult brain regulates behavior. *Science,* **2013**, *340*(6131), 449-453. [http://dx.doi.org/10.1126/science.1234152] [PMID: 23620046]

[78] Solis Herrera, A.; del Carmen Arias Esparza, M.; Solís-Arias, R.I.; Solís-Arias, P.E. The unexpected Capacity of Melanin to Dissociate the Water Molecule fills the Gap between the Life Before and After ATP. *Biomed. Res.,* **2010**,*21*(2), 221-224.

[79] Nickelsen, Kärin Otto Warburg´s first approach to Photosynthesis. *Photosynth. Res.,* **2007**, *92*, 109-120. [http://dx.doi.org/1007/s11120-0079163-3]

[80] Craig, P. Centennial history of the Carnegie Institution of Washington, The Department of Plant Biology. Cambridge University Press, Cambridge; 2005; Vol. IV: Available at: http://assets.cambridge. org/97805218/30829/excerpt/9780521830829_excerpt.htm

[81] Suchkov, S. The role of human-photosynthesis in predictive, preventive and personalized medicine. *EPMA J.,* **2014**, *5*(Supl. 1), A 146.

CHAPTER 2

The Cell Bioenergetics Pathways Beyond Glucose and Mitochondria and the Intrinsic Chemistry of Melanin

Arturo Solís Herrera*, María del Carmen Arias Esparza, Ruth I. Solís Arias, Paola E. Solís Arias and **Martha P. Solís Arias**

Human Photosynthesis® Study Center, López Velarde 108, Centro, Aguascalientes 20000, México

Abstract: What if glucose was not a source of energy?

Glucose is a very important metabolic intermediate and has a fundamental structural role. The function of the saccharides as source of biomass by excellence in our body is out of discussion, but not more than a source of energy. The unraveling of the intrinsic property of melanin to transform light energy into chemical energy through dissociation of water molecule radically changes our old belief that our body is able to obtain biomass and energy from the same molecule. A proof that our body and evolution has handled glucose (carbon chains) from the beginning of time can be found in the fact that the surfaces of all cells in all organisms are decorated with a dense and complex array of sugar chains. The biosynthesis, structure and function of glycans not only in the present but also throughout the evolution depend entirely on adequate levels of available chemical energy, so each and every single transformation that occurs in the glucose molecule can be carried out. Therefore, the form and function of living organisms are mainly determined by the distribution of energy, and in the same way also determines the geometry of the universe.

Keywords: Energy, Glucose, Glycobiology, Hydrogen, Melanin, Oxygen, Photosynthesis.

INTRODUCTION

Sugar or saccharides are essential components of all living things. The biology of saccharides (sugar chains or glycans) is studied by Glycobiology which is defined as the study of the structure, biosynthesis, and biology of saccharides. Glycobiology is fast emerging as a primary field of interest of biomolecular and biomedical research around the world. Once considered just supporting structures,

* **Corresponding author Arturo Solís Herrera:** Human Photosynthesis Study Center, Av. Aguascalientes Norte 607, Pulgas Pandas Sur, Aguascalientes, México; C.P. 20138; Tel/Fax: +524492517232; E-mail: comagua2000@yahoo.com

Arturo Solís Herrera (Ed.)

the sugars have now been widely recognized to be vital components in running (mainly structure) the complex biochemical machinery of life itself.

The study involves everything related (omics) to the structure, bio-synthesis, and biology of sugars and saccharides stemming from simple organic chemistry to molecular and cellular biology, enzymology and allied domains. Sugars exert a control and can influence almost every aspect of the cellular processes, ranging from cell-cell interactions, energy intermediates, adhesion mechanisms, growth factor signaling, blood clotting, receptor binding, regulating the activity of hormones in the blood, directing embryonic development, and also serving structural roles. Keeping in view the above mentioned properties, saccharides are considered a great source of energy. However, in light of our findings, the most intrinsic property of melanin is to transform visible and invisible light into chemical energy [1]. In this new paradigm, the role of sugars is being reconsidered with a new focus to understand the working of the intricate processes of life.

Background

Before sugars actually came into the limelight, it was commonly thought that life could begin with the biological information started from DNA and was ultimately translated into proteins which virtually controlled everything running in and around the cell. But now it has been demonstrated beyond doubt that simple sugars have their own complex language possibly more complex than DNA, starting by its unpredictability. These molecules continuously steer and guide many activities relevant to the proper functioning of a cell. In fact, many proteins undergo necessarily through post-translational modifications to form conjugated molecules with sugar (glycoproteins) to function properly.

Carbohydrates are structurally quite complex molecules. Their complexity can rival and overcome the size and complexity of DNA and proteins. Furthermore, their complexity increases as they come together to form a gamut of homo and hetero polymeric compounds inside the cell, controlling and influencing determinately a wide array of mechanisms. In accordance with Dr. Ajit Varki, the Glycobiology is the gathering of the traditional disciplines of carbohydrate chemistry and biochemistry with modern understanding of the cellular and molecular biology of glycans. Therefore, sugar molecules are not merely decorative elements serving simply as structural requirements in a cell. It is clear that their involvement in the intricate design of life from the beginning of the times, is far more crucial than that understood a few years ago.

Monosaccharides, defined as carbohydrates cannot be hydrolyzed by the cell into a simpler unit, are the basic structural units of glycans [2]. Glycans are at the center and thereby the focus of many disorders and diseases generating the possibility of exploiting them for therapeutic and diagnostic purposes. There are many biochemical pathways and diseases in which glycan are intricately involved in spite of their so far considered merely structure role. There are known and unknown ways in which sugars can affect the physiological conditions.

Glycome

The glycome refers to the current sugar population pattern in and out the cell, which can be in free or in conjugated forms, such as glycoproteins, glycolipids *etc.* They are not the same when are temporally observed, because glycans are continuously changing from time to time, making them more challenging. It should be considered that if a molecule is constantly changing, then its energy demands are too constant, even for the first time when it requires power to get activated.

A complete, functional glycan is quite complex. Unlike that of proteins whose composition can be derived from genetic code, the structure and conformation of glycans cannot be predicted easily. This makes it all the more challenging to study how exactly their production steps take place and are regulated. Furthermore, the production and regulation of glycans, as any other molecule in our body, undoubtedly need chemical energy.

Another very interesting aspect of glycan research is to analyze the biological importance of protein glycosylation. More than half of the protein molecules found in the human body have sugars conjugated to them. The process has to be thoroughly understood to determine at the most, the specific structures of N- and O-linked carbohydrates.

The scope of Glycobiology research and the expected possibilities are immense and even more when the approach can take in account that glucose is not a source of energy, but a biomass.

The Central Paradigm and Melanin

The central paradigm in modern molecular biology is that biological information flows from DNA to RNA to protein. There is often a tendency to assume the following extension of the central paradigm:

DNA → RNA → Protein → Cell → Organism

However, taking in account the unsuspected capacity of melanin to dissociate water molecule, the central paradigm could be changed in this way:

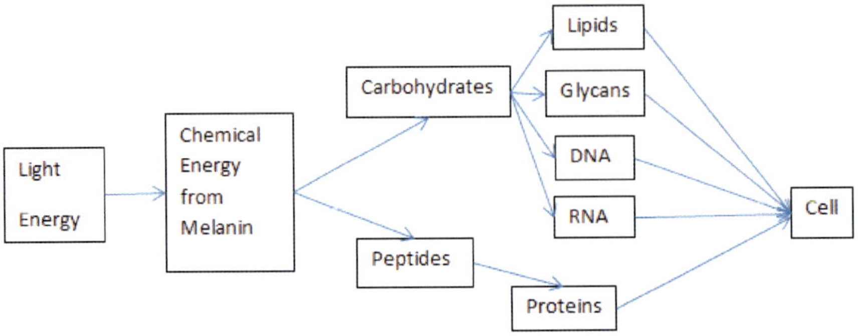

Presently, creating a cell requires two other major classes of molecules: lipids and carbohydrates. These molecules can serve as intermediates in the driving (not generation) of the fundamental chemical energy, as signaling molecules, or as structural components. The structural roles of carbohydrates become particularly important in constructing complex multicellular organs and thereby organisms, requiring interactions of cells between them and with the surrounding matrix. Monosaccharides significantly generate more linkage variation than amino acids or nucleotides [3].

All cells and many macromolecules that occur in nature carry a dense and complex array of covalently attached sugar chains (called oligosaccharides or glycans). In some instances, these glycans can also be free-standing entities. Since most glycans are on the outer surface of cellular and secreted macromolecules, they are in a position to modulate or mediate a wide variety of events in a cell-cell and cell-matrix interactions that are crucial to the development and function of a complex multicellular organism. They are also in a strategic position to mediate interactions between organisms, for instance, host and parasite. In addition, the simple, but highly dynamic protein-bound glycans are relatively abundant in the cell nucleus and cytoplasm, where they appear to serve as regulatory switches [4].

Neuron like any cell seems to use glucose as the primary source of energy, but this is a mistake. When plasmatic glucose levels decrease by half from a normal fasting value (to about 2-3 mM), this shows some cognitive impairment and, glucose levels below 1 mM shows mental confusion, and coma may result from sustained glucose deprivation. Therefore, it is an apparent firm belief that glucose is an essential energy source for the adult human brain. However, the signs and symptoms attributable to low levels of blood glucose highlight the delicate balance that exists between energy and biomass. Although our knowledge about the critical role of glucose metabolism in the maintenance of high level brain function has grown considerably in the recent years, the various factors that regulate glucose uptake and utilization in the CNS are not well understood.

The regulation of glucose metabolism in humans is a complex process that proceeds bi-directional, connecting the brain and peripheral tissues. Peripheral tissues, including muscle, fat, pancreas, and liver, are responsible for direct control over the transport, synthesis, storage, and metabolism of glucose [5].

Clinical Contradictory Data

1. - Injury to brain tissue in stroke is much more extensive in patients with hyperglycemia or diabetes as compared to patients with normal levels of blood glucose. Our comment about this is: If glucose is the primary source of energy, then the cell or the body must work, function, resist, or restore, much better conceptually, biomass and energy cannot be separated easily with higher levels of energy. The exploratory therapeutic administration of ATP, its precursors, its analogs and derivatives; in cases of neurologic diseases of traumatic or bleeding origin has been unsuccessful; even more the evolution of the damage is worst [6], it has been shown that treatment with exogenous ATP in experimental models of ischemic stroke is associated with exaggerated ischemic outcomes and dangerous systemic effects.

Constant darkness, a component of hibernation, increases AMP levels in blood [7], which is congruent with our experimental findings that when ATP is degraded to ADP or AMP energy is absorbed, not freed. Injection of exogenous AMP to mice and rats induces hypothermia (energy is absorbed) and hibernation-like condition. However, AMP-induced mild hypothermia in non-hibernating species like rats, failed to protect the brain from stroke because of multiple side effects, including hypotension and hyperglycemia [8].

ATP, ADP, and AMP are compounds with thermodynamic instability and kinetic stability. High levels of AMP can be interpreted as those chemical energy levels that are not adequate to compensate for the activation energy required for adding phosphate groups, because although the steps of AMP to ADP and ATP release energy, in any way requires activation energy. In other words, high levels of AMP means decreased levels of energy chemistry by whatever reason.

2. - Tardive dyskinesia. A serious movement disorder induced by neuroleptic drugs occurs more frequently in patients with elevated blood glucose levels. The question is again this: How is it possible that higher level of glucose and therefore available energy means more anatomic or functional impairment?

Theoretically, glucose is delivered to the brain by bloodstream and subsequently, it is taken up into cells and undergoes glycolytic breakdown to adenosine triphosphate (ATP) and pyruvate under aerobic conditions; this is: oxygen available as electrons acceptor so mitochondrial functions are normal. Pyruvate is transformed to acetyl-CoA *via* the tricarboxylic acid (TCA) cycle to generate ATP and reducing equivalents. According to the above statement, it is not possible to explain the deleterious effects of elevated blood glucose levels, it seems paradoxical. But if for a moment we do not consider glucose as a source of energy and think that it is the chemical energy which is required by cells, then the role of melanin becomes quite significant towards the whole body which transforms visible and invisible light into free chemical energy. Author's assistance required to verify this sentence. If glucose is the only source of carbon chains, *i.e.* the biomass, and the available chemical energy is not adequate then cell and the body cannot perform well which they would have normally performed with glucose availability. Author's assistance required to verify this sentence. Cell biology depends on a delicate balance finely tuned over millions of years by nature between mass and energy, as it is the geometry of the universe. If any of the components is not sufficient or both are altered, it results into a complete failure. It is widely accepted that a continuous supply of both oxygen and glucose is vital for normal metabolic function of the brain, which includes biochemical synthesis, normal synaptic transmission, maintenance of transmembrane ionic gradients, and cellular form and integrity.

However, if we take into account the intrinsic property of melanin to transform light energy into chemical energy throughout water dissociation, then the role of oxygen is not so essential, because each cell can produce oxygen at its own. On the contrary, molecular hydrogen (H_2) is a crucial reducing factor. The dissociation of water molecule explains well how the energy needs of the neurons are supplied.

3. - A number of neurologic and psychiatric illnesses are associated with alterations in energy metabolism in both the brain and peripheral tissues. The question arises if energy metabolism means glucose metabolism? Certainly not, even conceptually, biomass and energy cannot be separated easily. Glucose is mainly a source of building blocks, *e.g.* adenosine, ATP, nucleic acids; glycans, and so on, but energy is different.

4. - The content of ATP in fresh retinas obtained from control in diabetic rats shows 13.7 ±3 and 15.1 ±2, respectively. Data are the mean ± SD, nmol/retina [9]. Therefore, there is no doubt, that glucose is a source of biomass, and in this case of ATP, however the role of ATP must be reconsidered in light of our findings of the human photosynthesis or the hitherto unknown ability of melanin to split the water molecule and then produce chemical energy.

The measurements of lactate, pyruvate, and ATP, whose concentrations in rat retinal cells are known to be altered during a hypoxic episode [10], had not changes in the study of Diederen and Starns [2] by hyperglycemia, either in the *in vitro* incubations of retinas and cells or in the *in vivo* diabetic model. Thus, although blocking mitochondrial oxygen consumption leads to an increase in the content of retinal lactate, a decrease in the concentration of retinal ATP, and a decrease in the ratio of NAD-to-NADH8 (Winkler BS, unpublished results, 2003), none of these changes were observed under hyperglycemic conditions.

We must rethink that studies of glucose metabolism in the brain are limited, because the complex, integrating functions of the brain can only be studied in an intact, functioning brain in a conscious individual; whereas individual properties of brain cells, cell to cell interactions and mechanisms are most readily evaluated *in vitro* under controlled conditions, therefore, as in any other biological event translated to the test tube, the variables under study are irremissibly transformed into discreet variables.

Studies with labeled 2-deoxy-D-glcuose (DG) or 2-fluorodeoxyglucose (FDG) represent only one aspect of the brain function, identifying the pathways and magnitude of functional metabolic activities; the underlying contributions of different metabolic pathways, different cells and cells types in the brain are not identified and quantified, and the character of the energy-requiring processes are not determined.

The metabolic trafficking to sustain function has many heterogeneous interactions which have the consequence that imaging of overall brain metabolism cannot provide a picture of glucose metabolism at the cellular level [1].

Apart from the difficulty that these methods provide, no direct information about metabolic activities in the brain functioning *in vivo*, they are almost all encumbered with potential methodological problems as glycans in general.

Immunohistochemical studies of enzymes and substrate carriers in intact brain tissue have given quite useful information, however; they only provide information about the amount of enzyme or transporter protein, but not about the dynamic condition-dependent activity of the enzyme, or the transporter.

Well-differentiated primary cultures of neurons and astrocytes have similar rates of oxidative metabolism and similar contents of adenine nucleotides as the brain *in vivo* (Silver and Erecinska, 1997). However, tissue culture methodology has the potential source of error that the cultured cells may differ in metabolic characteristics from their in *vivo* counterparts.

Subcellular fractions obtained by dissociation of intact brain tissue followed by gradient centrifugation (*e.g.*, synaptosomes, mitochondria); have rendered ischemic (*i.e.*, exposed to severe energy failure and accompanying autolytic processes from which they may never fully recover).

Transit of glucose from blood across the capillary endothelium and ultimately into brain cells requires the action of several isoforms of the glucose transporter family; therefore that means more energy requirement. Glucose transporters GLUT1 and GLuT3 immunostaining increase in abundance in a region-specific manner following chronic seizures (Gronlund *et al.*, 1996) which is coherent with our finding that glucose is a source of biomass, but not of energy. Supposedly, if glucose was the source of energy then it must be increased before and not after seizures, because more neuron activity requires more energy, however, once tissues are damaged due to abnormal activity, then building blocks are required for repair and/or the replenishment of diverse biomolecules, thereby only glucose is the source of biomass, and of a highly complex biomass, besides complexity which means energy expenditure.

Studies of glycans lagged far behind those of other major class of molecules. This was in large part due to their inherent structural complexity, the difficulty in easily determining their sequence, and the fact that their biosynthesis could not be directly predicted from the DNA template.

The surfaces of most types of cells are heavily decorated with different kinds of glycoconjugates, therefore are effectively covered with a dense coating of sugars, the so-called glycocalix.

Nucleotides and proteins are linear polymers that can each have only one basic type of linkage. In contrast, each monosaccharide can theoretically generate β or α linkage to any of the several positions on another monosaccharide chain or to another type of molecule. Thus, it has been pointed out that although three nucleotide bases or amino acids can only generate six variations, three hexoses could produce (depending on which factors are considered) anywhere between 1.056 to 27.648 unique trisaccharides. As the number of units in the polymer increases, this difference in complexity becomes even greater. For example, a hexasaccharide with six hexoses could have more than 1 trillion possible combinations. We must keep in mind that any change requires chemical energy expenditure.

The common monosaccharides units found in higher animals′ oligosaccharides or glyconjugates [11] are:

1. - Sialic acids (Sia) of which the most common is N-acetyl neuraminic acid.

2. - Hexoses: Six- carbon neutral sugars, including glucose (Glc), galactose (Gal), mannose (Man).
3. - Hexosamines: Hexoses with an amino group at the 2-position, which can be either free or, more commonly, N-acetylated; N-acetylglucosamine (glcNAc) and N-acetylgalactosamine (GalNAc).
4. - Deoxyhexoses: Six-carbon neutral sugar without the hydroxyl group at the 6-position, fucose (Fuc).
5. - Pentoses: Five-carbon sugar, xylose (Xyl).
6. - Uronic Acids: Hexose with a negatively charged carboxylate at the 6-position, glucuronic acid (GlcA) and iduronic acid (IdA).

Most well-characterized pathways for the biosynthesis of different classes of glycans occur within the ER-Golgi-plasma lemma pathway and its other ramifications. Thus, newly synthesized proteins or lipids originating from the ER are either co-translational or post-translational modified with sugar chains at various stages in their itinerary toward their final destinations. A variety of factors determine the final outcome of glycosylation reactions that occur within the ER and the Golgi apparatus and every step needs energy expenditure, otherwise, it occurs in a random manner, which is not compatible with life.

Until the mid-1980s, a firm belief was that glycoconjugates, such as glycoprotein and glycolipids, occur exclusively on the outer surface of cells, on the internal (luminal) surface of intracellular organelles, and on secreted molecules. However, the cytosol and nucleus are topologically semi continuous, because the existence of nuclear pores was assumed to be devoid of glycosylation capacity. And in the last two decades, it has become clear that certain types of glycoconjugates are synthesized (therefore requires energy expenditure) and reside within the cytosol and nucleus, GlcNAc may well be numerically, the most common type of glycoconjugates in many cells. Therefore, nuclear and cytoplasmic Glycosylation is common [12].

In naturally occurring glycoconjugates, the portion of the molecule comprising the glycans can be carried to a great extent from being very minor in amount to a dominant component. It is striking that sugar chains make up a substantial portion of the mass of most of the glycoconjugates. The surfaces of most type of cells, which are heavily decorated with different kinds of glycoconjugates are effectively covered with a dense coating of sugars, giving rise to the so-called glycocalix. This cell surface structure was first observed by electron microscopists many years ago as an external anionic layer to the plasma lemma, which could be decorated with polycationic reagents like cationized ferritin.

The common classes of oligosaccharides found on eukaryotic cells are primarily defined according to the nature of the linkage (core) regions to the glycone (protein or lipid). A glycoprotein is a glycoconjugate in which a protein carries one or more oligosaccharide chains covalently attached to a polypeptide backbone, usually *via* N- or O-linkages. On the contrary, a proteoglycan is a glycoconjugate having one or more covalently attached glycosaminoglycan chains. The distinction from a glycoprotein is otherwise arbitrary, since some polypeptides can carry both glycosaminoglycan chains and N- or O-linked chains.

Mucin is a normal component of several interphases, *i.e.* tears film; its apparent function is to maintain the tears film attached over the surface of the eye, which is a large glycoprotein that carries many O-glycans that are often closely spaced, clustered-like.

The glycosphingolipid, as cerebrosides, galactocerebrosides, gangliosides, globosides, and glycophosphosphingolipids; are oligosaccharides usually attached *via* glucose or galactose to the terminal primary hydroxyl group of the lipid moiety ceramide, which is itself composed of a long chain base (*i.e.* sphingosine) and a fatty acid. The glycosphingolipids, often called glycolipid, can be neutral or anionic. A ganglioside is an anionic glycolipid containing one or more residues of sialic acid. These represent the most common classes of glycans reported in eukaryotic cells. There are several other less common types found on both side of the cell membrane of animal cells.

Melanin as Unsuspected Source of Chemical Energy

Melanin is a molecule with unique features in many ways. It is well known for more than 200 years that squid ink is melanin, and found in good condition in the sacks of the squid fossilized who died 170 million years ago [13]. It is a deep rooted belief that the main function of the melanin in humans is a simple built-in sunscreen that protects us from the UV radiations of the Sun. Sir Everard Home, in 1820; was the first to propose this concept, and it has not changed very much in the intervening 195 years [14]. Body absorption of visible or infrared radiation will only heat the body (Morison, 1985).

The great pigments of living things are hemoglobin, chlorophylls, carotenoids, flavonoids, and melanin. The prosthetic group of the first four substances is orderly arrays of single and double bounds. When melanin is isolated from living tissue, it produces an insoluble and amorphous mud that defies analysis by classical techniques [15]. The unsuspected intrinsic capacity of melanin transforms light energy into chemical energy through the water molecule dissociation detected during a descriptive, observational study about the three main causes of blindness in the world, which are: glaucoma, diabetes and macular degeneration by age.

Melanin is thousand times more efficient than chlorophyll to transform light energy into chemical energy, for various reasons; among them we have chlorophyll is only capable of using the ends of the visible spectrum (blue and red) to dissociate the water molecule in an irreversible manner, and can be represented in the following way:

$$2H_2O \rightarrow 2H_2 + O_2$$

While melanin, the darker substance that is known, is able to capture the energy of the electromagnetic spectrum in its entirety, this is from gamma rays to radio waves; and on the other hand, it is not only melanin which is able to dissociate the water molecule, separating its component molecules; it is hydrogen and oxygen; but it is also able to withstand the backlash: *i.e.*: melanin which is capable of reforming the water molecule, which means that for every two molecules of reformed water, 4 high-energy electrons are generated. The reaction that takes place inside the melanin is exemplified in the following manner:

$$2H_2O \leftrightarrow 2H_2 + O_2 + 4e^-$$

This chemical energy is released by the melanin in the form of diatomic hydrogen and high energy electrons; given that melanin is contained in melanosomes located, in turn; in the cytosol, and specifically in the perikaryon or perinuclear space and it is symmetrically distributed in all directions (Fig. **1**), the flow of this chemical energy reaches until the last corner of the cell. It should be considered that diatomic hydrogen does not combine with water.

Energy Definition

Energy could be defined as everything that produces a change. Therefore, any transformation that occurs to atoms or chains of carbon, this is: shortening, lengthening, torque, or combination requires undoubtedly the expenditure of chemical energy by the cell. The behavior of the glycans in the body somehow reflects the energy that gives rise to them.

Thus, the low-molecular-weight sugar nucleotides that act as donors for most of the biosynthetic steps are made in the cytosol. After that they are specifically transported into the lumen of the organelles, which are processes that require certain level of chemical energy to carry out, in the right way.

It has become clear that certain types of glycoconjugates are synthesized and reside within the cytosol and nucleus, but the nuclear synthesis of every single molecule requires chemical energy expenditure, not only as initial step instead along the whole process, however, the nucleus has neither mitochondria nor ATP. This apparent failure could be explained by the fact that melanosomes tend to form a kind of envelop around the nucleus. The energy that emanates from the melanin is released in a symmetrical form in all directions, like growing spheres. This chemical energy emerges from melanin in the form of spheres of alternating nature, *i.e.* molecular hydrogen, the energy carrier by excellence in the universe and flow of high energy electrons. This continuous flowing of alternating energy spheres tend to coalesce and overlap with each other, and due to the fact that melanosomes form a type of envelope around the nucleus, therefore the emanated energy spheres bring together to form a center of high energy precisely in the core of the cell nucleus (Fig. **1**).

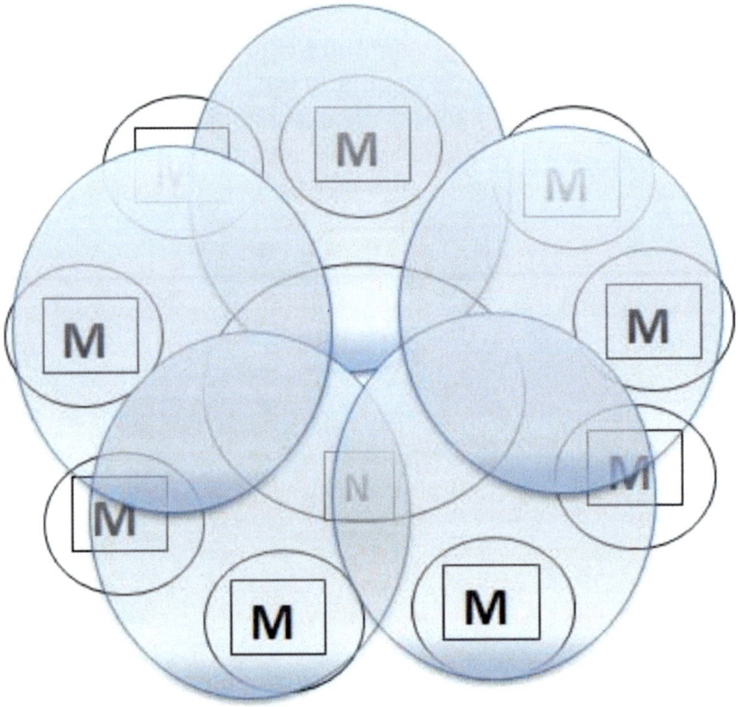

Fig. (1). The nucleus (N) surrounded by melanosomes containing melanin (M). The zone of the nucleus where the spheres of growing alternating energy (blue) formed by water/gas/water continuous process are overlapping and coalescing forming areas of highest chemical energy concentration, *i.e.* the center of the nucleus in the eukaryotic cell; this explains how the nucleus energy requirements are fulfilled, in spite of the normal absence of mitochondria and ATP in it.

That cytoplasmic glycosylation requires available chemical energy as evident by the fact that the active sites of the relevant glycosyltransferases face the cytosol. Until the mid-1980s, a commonly stated dogma was that glycoprotein and glycolipids (glycoconjugates), occur exclusively on the outer surface of the cells, on the internal (luminal) surface of intracellular organelles, and on the secreted molecules. The cytosol and nucleus appear to lack this capacity. However, in the last two decades, it has become clear that certain types of glycoconjugates are synthesized (therefore need chemical energy expenditure) and reside within the cytosol and nucleus. The fact that the major form of glycosylation (O-linked GlcNAc) was missed by so many investigators for so long emphasizes the relatively unexplored state of the whole field of Glycobiology.

One of the most fascinating and yet frustrating aspect of protein glycosylation is the phenomenon of micro-heterogeneity. This term indicates that at any given glycosylation site on a given protein synthesized by a particular cell type, requires chemical energy expenditure, we have a range of variations that can be found in the precise structure of glycan. Even the extent of this heterogeneity can vary considerably from glycosylation site to glycosylation site, from protein to protein, and even from cell type to cell type. Thus, a given glycoprotein can exist in numerous glycoforms, each effectively being a distinct molecular species. Mechanistically, this heterogeneity might be explained by the rapidity with which multiple, sequential, partially competitive glycosylation reactions must take place in the Golgi apparatus through which the newly synthesized glycoprotein is passing. Every single step in the Golgi apparatus requires chemical energy, as in any other place in and outside the cell. ATP pathway is not enough to supply energy to all these reactions, besides the fact that when ATP is downgraded to ADP, energy would be absorbed.

From the practical point of view, micro heterogeneity explains the anomalous behavior of glycoproteins in various forms of chromatography and makes the complete structural analysis of most of the glycoproteins a difficult task. From a functional point of view, the meaning of this heterogeneity remains unclear. There are a lot of functions of glycans, but if we do not consider the energy generation then the complexity will be reduced to some extent which is due to their completely functional role.

Like all components of living cells, glycans are constantly being created and degraded [16] and therefore energy also needs to be constant. Degradation is mediated by enzymes that also require chemical energy and cleave sugar chains either at the outer (no reducing) terminal end (exoglycosidase) or internally (endoglycosidases). Some outer units can also be removed, also through a chemical energy requiring process; and then reattached (energy is needs) without degradation of the underlying chain. The final complete degradation of most glycans is generally carried out by a series of glycosidase in the lysosome; however, in any case chemical energy expenditure is required. Once there are broken down, their individual unit, monosaccharides is then typically exported from the lysosome, and this exportation also requires chemical energy expenditure into cytosol so that they can be reutilized again. In contrast to the relatively slow turnover of glycans derived from the ER-Golgi pathway, glycans of the nucleus and cytoplasm may be more dynamic and rapidly turned over, but in one way or another, energy requirements are never-ending, because not only energy is required for the synthesis of the glycans, but also to maintain the form and function.

Remarkably, little is known about the evolution of glycosylation. There are clearly and unique sharing features of glycosylation in different kingdoms and taxa, and among animals, an increasing complexity is often observed in higher forms. Many natural bioactive molecules are glycoconjugates. For this reason alone, glycobiology and carbohydrate chemistry have increasing importance in the modern biology.

Glucose: The Universal Building Block

Glucose is a crucial building-block for cartilage function *in vivo* as it is for many other tissues and organs. Future progress in dealing with degenerative joint disorders, such as OA (Osteoarthritis), ACD, and related joint disorders will be highly dependent on a better understanding of the unique nutritional requirements of chondrocytes [17].

The requirements of energy, structural elements, and metabolic intermediates, are ever changing depending if the cells are involved in different activities, such as division, proliferation, differentiation, and apoptosis.

Glucose is an example of a principal structural substrate but not of metabolic fuel. Nutrition is critical to the survival of all prokaryotes and eukaryotes; however nutrition itself must be divided into building blocks sources, such as glucose and photosynthesis as primary energy source.

Glucose is the most abundant monosaccharide on the planet and is the principal carbon chain source of the cell, but the glucose itself cannot be the only energy source at the same time. Glucose is an important carbon currency on land and in the oceans where plants and phytoplankton actively fix carbon dioxide into carbohydrates. At both unicellular and multicellular levels, the quantity of available glucose can fluctuate considerably and living organisms must be able to sense the amount of glucose available to vital tissues and organs and respond appropriately by adjusting their metabolic rate, however it has a very slow response, demonstrated by the fact that with no meal, human body can survive up to three months approximately. The water as a source of energy needs to be much more efficient, and could be appreciated by the fact that with no water, human being dies usually in three days.

The Regularity in Living Things

The storage site for polyphosphates may represent the first known universal organelle. Polyphosphate storage organelle is present in all three domains of life — bacteria, archaea and the eukaryotes, which include animals, plants and fungi. The polyphosphate storage structure in at least two bacterial species was physically, chemically and functionally the same as an organelle called an acidocalcisome found in many single-celled eukaryotes.

Nowadays, both DNA and proteins require each other to form. It is more difficult to explain how each entity formed independently, it is more logical to think that both were formed at the same time as the RNA which can store information like DNA, serve as an enzyme like proteins, and help create both DNA and proteins, because they are more efficient. RNA still exists and performs several functions in organisms, including acting as an on-off switch for some genes. But clarify that it was first DNA, RNA, or proteins does not make much sense, because the initial spark common to all them was the constant presence of available chemical energy. There, the different forms of life evolved in their own way depending on the environment, this is: in some forms of life appearing first some proteins, in other DNA, and other RNA, but even there regularly exists, and such uniformity would start from the chemical energy available both in nature and in features. For example: Energy coming from melanin is always in the form of hydrogen diatomic and high-energy electrons.

The genetic code is essentially the same in all known living organisms, and several core functions, such as gene transcription and chemical energy generation which tend to be conserved across various taxa [18]. But it remains a mystery what dictates the uniformity and the long lasting conservation of characters along the evolution and furthermore into the various forms of life. But from the beginning of time to date, the chemical energy has been similar in each and every one of its characteristics, *i.e.*: the levels of hydrogen diatomic and high-energy electrons; this has continued and thereby confirmed by the uniformity at the cellular level, because all cells have an amazing similarity in terms of organelles, even in size. Then, if uniformity starts from the first spark of life, from chemical energy of dissociation and the subsequent reforming of the water molecule, then the regularity of the cells is also determined by the intrinsic regularity, by which melanin releases chemical energy. This was expected since melanin has not evolved, maybe because it has the ideal characteristics suitable to be the perfect candidate to transform light energy into chemical energy from the beginning of time. As the geometry of the universe is particular for the generation and distribution of energy, similarly, the shape and function of cells and therefore living beings are mainly determined by the generation and distribution of energy.

The variations in the secondary characteristics of the organism if for *e.g.* flying, walking, if it creeps or if it is relatively fixed as plants, would be conditioned by the environment. It is doubted that life is originated in one place in particular; it is more plausible that it has evolved in various places at the same time, and the fundamental characteristics were similar due to identical type of chemical energy that resulted in the emergence of life.

Melanin is a more stable substance that is known to display the same characteristics till date, as shown 170 million years ago [19]. The main function of melanin before and now is identical: the transformation of light energy into highly ordered chemical energy through the dissociation of the water molecule. What undoubtedly required evolution were the members of the cell that they ended up being very similar in all the kingdoms of mother nature, and so much so that melanin is in all of them, in one way or another and predictably with the same function: energy production. In our opinion, the first organelle that appeared was probably the cell membrane that surrounds the granules of melanin and hence, gradually the other organelles appeared along billions of years of evolution. By other hand, Melanin is not free-floating in the cell cytoplasm but always surrounded by a bilayer lipid, which allows its main products: molecular hydrogen and high-energy electrons to flow properly.

Each cells of the body receive the sunlight which can be displayed through different colors of the skin. In places where there is plenty of light, the amount of melanin in the skin is greater, thus regulates the amount of light which penetrates inside the body, because it should be in the right amount neither more nor less. Thus, deep into the body, bone requires to capture light energy to transform it into chemical energy. In places where the amount of light is reduced, for example in cold countries, the amount of melanin in the skin is significantly less to allow a greater light to pass.

Glucose and Energy

Glucose has long been time considered (wrongly) as the main energy provider in our organism; supposedly, the main chemical energy provided to the body is through glucose, therefore it has a primary role in our organism, a concept that is repeated over and over again in any textbook on the subject. Furthermore, gradually other additional roles about glucose and other sugars in the body were uncovered. Sugar molecules dangle from many of the body´s protein and fat molecules with dramatic consequences. Sugar attached to a protein changes the protein´s shape and function; when protein binds to lipids in cell membranes, sugars alter the way through which cells recognize each other. Nevertheless, the actual intracellular pathway leading to increased genes transcription in response to carbohydrate feeding is poorly understood [20].

Kidney Cortex and Chemical Energy

The glomerular capillary tuft is a highly intricate and specialized microvascular bed that filters plasma water and solute to form urine. The mature glomerulus contains four cell types: Parietal epithelial cells that form Bowman's capsule, podocytes that cover the outermost layer of the glomerular filtration barrier, glycocalix-coated fenestrated endothelial cells that are in direct contact with blood, and mesangial cells that sit between the capillary loops. Filtration begins only after the influx and organization of endothelial and mesangial cells in the developing glomerulus. Tightly coordinated movement and cross-talk between these cell types is required for the formation of a functional glomerular filtration barrier, and disruption of these processes has devastating consequences in early life [21].

Mesangial cells and their matrix form the central stalk of the glomerulus and are part of a functional unit interacting closely with endothelial cells and podocytes. Alterations in one cell type can produce changes in the others. The cytokines generated by mesangial cells, endothelial cells, and podocytes that tridirectionally and interactively influence cognate receptors on receiver cells are not fully defined. The existence of cytokine cross-talk seems very likely, given the observations that podocytes injury frequently results in mesangial cell proliferation, whereas mesangial cell injury leads to foot process fusion and proteinuria. Another potentially fruitful area of future research is the role of mesangial cells as local modulators of innate and adaptive immune responses [22]. The metabolic activity of kidney indicates that the cells that comprise it use substantial amounts of chemical energy to carry out their highly complex functions.

Adenosine, the Backbone of ATP

Under normal conditions, the intracellular and extracellular adenosine, a nucleoside that is composed of adenine and d-ribose, that play many important biological roles in addition to being components of DNA and RNA; is continuously generated, therefore it continuously needs energy. The intracellular production is mediated either by an intracellular 5-nucleotidase, which dephosphorylates AMP (Schubert *et al.*, 1979; Zimmermann *et al.*, 1998), or by hydrolysis of *S*-adenosyl-homocysteine (Broch and Ueland, 1980). Adenosine generated intracellularly is transported into the extracellular space, mainly *via* specific bi-directional transporters through facilitated diffusion that efficiently evens out the intra- and extracellular levels of adenosine. In some tissues (*e.g.*, kidney brush-border membranes) there is a concentrative nucleoside transport protein capable of maintaining high adenosine concentrations against a concentration gradient. It must be remembered that in tissue, some cells are net producers of adenosine and in these cells intracellular levels rise, whereas most cells are net eliminators of the nucleoside. Adenosine itself is a neurotransmitter that convey information, and the normal response of the cell depends on the chemical energy present inside it, is within the appropriate levels.

The de-phosphorylation of extracellular AMP to adenosine, mediated by ecto-5--nucleotidase, is the last step in the enzymatic chain that catalyzes the breakdown of extracellular adenine nucleotides, such as ATP, to adenosine. The entire catalytic pathway can be completed in a few hundred milliseconds, and the rate-limiting step seems to be the de-phosphorylation of AMP to adenosine by ecto-5_-nucleotidase (Dunwiddie *et al.*, 1997a).

When given in very high amounts, adenosine can affect intracellular nucleotide pools and even provide a source of metabolizable energy. In addition, it has been reported very recently that the human growth hormone secretagogue receptor (GHS-R) also accepts adenosine as a highly potent endogenous agonist, in addition to the endogenous peptide GHS-R agonist, ghrelin (Smith *et al.*, 2000; Tullin *et al.*, 2000). However, most effects of adenosine are due to activation of adenosine receptors [23].

Adenosine can also be released into the extracellular space after application of specific neurotransmitter ligands. Glutamatergic agonists, such as NMDA or kainate, dose dependently increase adenosine levels (Carswell *et al.*, 1997; Delaney *et al.*, 1998). Activation of NMDA receptors seems to release adenosine itself rather than a precursor (Manzoni *et al.*, 1994; Harvey and Lacey, 1997). Dopamine D1 receptors enhance adenosine release *via* an NMDA receptor-dependent increase in extracellular adenosine levels (Harvey and Lacey, 1997), but dopamine depletion causes no significant changes in the extracellular levels of striatal adenosine as measured by *in vivo* microanalysis (Ballarin *et al.*, 1987).

Thus, dopaminergic input may be important to transiently elevate adenosine but not so important in maintaining a basal level of the nucleoside. Nitric oxide can also control basal levels of endogenous adenosine *in vivo* (Fischer *et al.*, 1995; Delaney *et al.*, 1998) as well as *in vitro* (Fallahi *et al.*, 1996).

Another potential source of extracellular adenosine is cAMP, which can be released from neurons and converted by extracellular phosphodiesterases into AMP, and thereafter by an ecto-5_-nucleotidase to adenosine.

Functional evidence for a relevant role of this pathway has been obtained in the ventral tegmental area and hippocampus (Bonci and Williams, 1996; Brundege *et al.*, 1997; Dunwiddie *et al.*, 1997a,b). However, to provide a physiologically important adenosine release, it seems that multiple cells must release cAMP over a prolonged period (Brundege *et al.*, 1997).

The levels of adenosine, at least in the basal forebrain, striatum, hippocampus, and thalamus, are higher during wakefulness than sleep (Huston *et al.*, 1996; Porkka-Heiskanen *et al.*, 1997). This fact is of greater importance that probably whenever intracellular levels of adenine nucleotides fall as a result of excessive energy use, the intracellular levels of adenosine will rise dramatically (Rudolphi *et al.*, 1992). For example, following hypoxia (Zetterström *et al.*, 1982), ischemia (Berne *et al.*, 1974), or electrical stimulation (Pull and McIlwain, 1972), there is a decrease of intracellular ATP, accompanied by an accumulation of 5_-AMP and subsequently adenosine. The nucleoside is then transported into the extracellular space. When the intracellular level of adenosine is very high, adenosine simply diffuses out of cells.

Furthermore, in general, the higher the number of receptors the more potent and/or efficacious will be the agonist. Thus, the rather low levels of endogenous adenosine present under basal physiological conditions have the potential of activating receptors where they are abundant, but not where they are sparse (Kenakin, 1993, 1995; Svenningsson *et al.*, 1999c; Kull *et al.*, 2000b; Fredholm *et al.*, 2001).

Adenosine is believed to play modulatory roles in a variety of tissues and physiological circumstances. Through information mainly.

Adenosine is not primarily released in a transmitter- or hormone-like fashion, but instead it appears to be formed by group of cells as part of a response, *e.g.*, to challenges in energy metabolism. It can also be formed by breakdown of ATP released by cells either in a regulated fashion or in response to massive trauma. Adenosine is therefore likely to act in concert with several other messengers (transmitters, hormones, growth factors, autacoids).

Adenosine can activate phospholipase C *via* adenosine A1 receptors and a Gi-dependent mechanism. Interestingly, adenosine acts synergistically with nucleotides, such as ATP or UTP (Gerwins and Fredholm, 1992a), histamine (Dickenson and Hill, 1993), or with bradykinin (Gerwins and Fredholm, 1992b). ATP, histamine, and bradykinin instead act *via* Gq/11.

Given that ATP is rapidly broken down to adenosine and that ATP and adenosine act synergistically, we may be dealing with a quite important mechanism, biologically. We still do not understand the factors that are important in regulating adenosine A1 and A3 receptors under physiological and pathophysiological conditions.

CONCLUSION

The different models of metabolism-first, gene-first, RNA-first, ATP-first, are now displaced by the scenario of melanin-first. Perhaps life did not begin on Earth at all, but was brought here from elsewhere in space, but not like a notion known as panspermia, instead a melanin in space. Melanin is the most primitive and universal pigment in living organisms. In fact, civilizations have risen and fallen due to perceptions and misperceptions, concerning its nature and significance.

It is an observed fact that greater melanin means less mitochondrial number (until 83% less, and respiration down 30%), and *vice versa.* These findings reflect significant differences in metabolism between white and blacks.

In the United States, large morbidity and mortality differences are observed between Blacks and Whites [24]. Income, education, and several known risk factors, such as smoking, diabetes, and increased blood pressure, cholesterol, body mass index, and alcohol intake are known to contribute only partly to the excess of those diseases in the Black population. However, these factors must be reconsidered in light of unsuspected intrinsic property of melanin to split the water molecule.

For instance, one cup of alcohol decreases the transduction light-chemical energy for almost two weeks, due to ethanol should form covalent bonds with corresponding receptors. On the contrary, cold weather has a significant impact on the process efficiency, like in plants. Black bodies resist cold significantly less than Whites, probably by the difference in mitochondria number.

Also, comorbidity of HIV and tuberculosis is a common phenomenon that is more frequent in Blacks; therefore, analysis of data must be taken into account when considering the difference in lungs size, which is smallest in Blacks and larger in Whites.

Rates of death for suicide and melanoma were lower among Black as compared to their White counterparts, as were those for coronary heart diseases after adjustment for income [25]. However, higher levels of chemical energy due to more melanin content, means less depression; for instance suicide rates are significantly less in sunny countries. In regards to melanoma, at present, Black and White skins are burn in a similar way, but the presence of melanin contributes to a higher available chemical energy that means to improve and fasten tissue recovery.

Furthermore, Black veterans who present with acute coronary insufficiency are less likely than Whites to have significant coronary obstruction. Current understanding of coronary disease does not provide an explanation for these differences [26]. The clue could be that higher melanin content means higher oxygen tissue levels.

Nevertheless, high oxygen levels have secondary importance as compared to high diatomic hydrogen levels, because hydrogen is the best energy carrier in the universe. Highest hydrogen levels imply more available chemical energy and more healthy tissues. Therefore, this is probably the best explanation to life expectancy in Blacks, having advantage in chronic liver disease, Alzheimer's disease, chronic lower respiratory diseases, and suicide [27].

The differences between the diseases that are observed according to the amount of melanin in the body have a new factor that must be considered to advance knowledge. Melanin has gone from being a simple solar filter built-in to a molecule basic to bio-energy processes of the human body, similar to the chlorophyll in plants. You have to rethink about cell biology in front of the epistemological revolution which implies that the intrinsic property of melanin transforms light energy into chemical energy by means of dissociation of the water molecule.

HUMAN AND ANIMAL RIGHTS

No humans were used in this research. All animal research procedures followed were in accordance with the standards set forth in the eighth edition of Guide for the Care and Use of Laboratory Animals published by the National Academy of Sciences, The National Academies Press, Washington, D.C.

CONFLICT OF INTEREST

The author (editor) declares no conflict of interest, financial or otherwise.

ACKNOWLEDGEMENT

We thank the unconditional support of the Center for studies of Human Photosynthesis®, S.C.

REFERENCES

[1] Solis-Herrera, A.; Arias-Esparza, M.C.; Solis-Arias, R.I. The unexpected capacity of melanin to dissociate the water molecule fills the gap between the life before and after ATP. *Biomed. Res.*, **2010**, *21*(2), 224-226.

[2] Varki, A.; Manzi, A.E.; Freeze, H.H. Introduction: Preparation and analysis of glycoconjugates. In: *Current Protocols in Molecular Biology*; Ausubel, F.M., Ed.; Unit 17.0. Wiley: New York, **1996**; pp.

28-33.

[3] Rademacher, T.W.; Parekh, R.B.; Dwek, R.A. Glycobiology. *Annu. Rev. Biochem.,* **1988**, *57*, 785-838. [http://dx.doi.org/10.1146/annurev.bi.57.070188.004033] [PMID: 3052290]

[4] Varki, A.; Cummings, R.; Esko, J. *Essentials of Glycobiology*; Cold Spring Harbor Laboratory Press, **1999**.

[5] Dwyer, D.S. Editor in the book "*Glucose Metabolism in the Brain*", International Review of Neurobiology. Elsevier Science, USA, **2002**. *Vol. 51* 0074-7742/02. Available at: https://www.elsevier.com/books/book-series/international-review-of-neurobiology

[6] Zhang, M. L.; Wenjing, N.; Guangming, L.; Rehana, K.; Chen, J.; Zhang, F. ATP induces mild hypothermia in rats but has a strikingly detrimental impact on focal cerebral ischemia. *J. Cereb. Blood Flow Metab.,* **2013**, *33* [http://dx.doi.org/10.1038/jcbfm.2012.146]

[7] Zhang, J.; Kaasik, K.; Blackburn, M.R.; Lee, C.C. Constant darkness is a circadian metabolic signal in mammals. *Nature,* **2006**, *439*(7074), 340-343. [http://dx.doi.org/10.1038/nature04368] [PMID: 16421573]

[8] Zhang, F.; Wang, S.; Luo, Y.; Ji, X.; Nemoto, E.M.; Chen, J. When hypothermia meets hypotension and hyperglycemia: the diverse effects of adenosine 5′-monophosphate on cerebral ischemia in rats. *J. Cereb. Blood Flow Metab.,* **2009**, *29*(5), 1022-1034. [http://dx.doi.org/10.1038/jcbfm.2009.28] [PMID: 19319149]

[9] Diederen, R.M.H.; Starnes, C.A.; Berkowitz, B.A.; Winkler, B.S. Reexamining the hyperglycemic pseudohypoxia hypothesis of diabetic oculopathy. *Invest. Ophthalmol. Vis. Sci.,* **2006**, *47*(6), 2726-2731. [http://dx.doi.org/10.1167/iovs.06-0076] [PMID: 16723492]

[10] Winkler, B.S.; Arnold, M.J.; Brassell, M.A.; Sliter, D.R. Glucose dependence of glycolysis, hexose monophosphate shunt activity, energy status, and the polyol pathway in retinas isolated from normal (nondiabetic) rats. *Invest. Ophthalmol. Vis. Sci.,* **1997**, *38*(1), 62-71. [PMID: 9008631]

[11] Varki, A. Biological roles of oligosaccharides: all of the theories are correct. *Glycobiology,* **1993**, *3*(2), 97-130. [http://dx.doi.org/10.1093/glycob/3.2.97] [PMID: 8490246]

[12] Brockhausen, I. Pathways of O-glycan biosynthesis in cancer cells. *Biochim. Biophys. Acta,* **1999**, *1473*(1), 67-95. [http://dx.doi.org/10.1016/S0304-4165(99)00170-1] [PMID: 10580130]

[13] Hart, G.W. Dynamic O-linked glycosylation of nuclear and cytoskeletal proteins. *Annu. Rev. Biochem.,* **1997**, *66*, 315-335. [http://dx.doi.org/10.1146/annurev.biochem.66.1.315] [PMID: 9242909]

[14] Joseph, B.; Wassermann, N.; Vratsanos, Spyros. Erlanger, Bernard F. Photoregulation of biological activity by photochromic reagents, IV, A model for DDiurnal variation of enzymic activity. *PNNAS,* **1970**, *66*(3), 850-854. [http://dx.doi.org/10.1073/pnas.66.3.850]

[15] Sweeley, C.C. Glycosphingopidis: structure and function. *Pure Appl. Chem.,* **1989**, *61*(7), 1307-1312. [http://dx.doi.org/10.1351/pac198961071307]

[16] Glass, K.; Ito, S.; Wilby, P.R.; Sota, T.; Nakamura, A.; Bowers, C.R.; Vinther, J.; Dutta, S.; Summons, R.; Briggs, D.E.; Wakamatsu, K.; Simon, J.D. Direct chemical evidence for eumelanin pigment from the Jurassic period. *Proc. Natl. Acad. Sci. USA,* **2012**, *109*(26), 10218-10223. [http://dx.doi.org/10.1073/pnas.1118448109] [PMID: 22615359]

[17] Morison, W.L. What is the function of melanin? *Arch. Dermatol.,* **1985**, *121*(9), 1160-1163.

[18] Hill, H.Z. The function of melanin or six blind people examine an elephant. *BioEssays,* **1992**, *14*(1), 49-56.
[http://dx.doi.org/10.1002/bies.950140111] [PMID: 1546980]

[19] Sandhoff, K.; Kolter, T. Topology of glycosphingolipid degradation. *Trends Cell Biol.,* **1996**, *6*(3), 98-103.

[20] Mobasheri, A.; Bondy C.A.; Moley, K. Facilitative Glucose Transporters in Articular Chondrocytes. In the series Advances in Anatomy Embriology and Cell Biology published by Springer-Verlag: Berlin, Heidelberg, **2008**.

[21] Ito, Shoshuke Melanins seem to be everywhere in the body, but for what? *Pigment Cell Melanoma Res.,* **2009**, *22*, 12-13.
[http://dx.doi.org/10.1111/j.1755-148X.2008.00538.x]

[22] Ramos, I.; Gomes, F.; Koeller, C.M.; Saito, K.; Heise, N.; Masuda, H.; Docampo, R.; de Souza, W.; Machado, E.A.; Miranda, K. Acidocalcisomes as calcium- and polyphosphate-storage compartments during embryogenesis of the insect Rhodnius prolixus Stahl. *PLoS One,* **2011**, *6*(11), e27276.

[23] Ruiz, F.A.; Lea, C.R.; Oldfield, E.; Docampo, R. Human platelet dense granules contain polyphosphate and are similar to acidocalcisomes of bacteria and unicellular eukaryotes. *J. Biol. Chem.,* **2004**, *279*(43), 44250-44257. [November].
[http://dx.doi.org/10.1074/jbc.M406261200] [PMID: 15308650]

[24] Varki, A. Freeze, Hudson H. Gagneux, Pascal. Evolution of glycan diversity, Chapter 19 of the Book: Essentials of Glycobiology Cold Spring Harbor Laboratory Press: Cold Spring Harbor (NY), **2009**.

[25] Glass, K.; Ito, S.; Wilby, P.R.; Sota, T.; Nakamura, A.; Bowers, C.R.; Vinther, J.; Dutta, S.; Summons, R.; Briggs, D.E.; Wakamatsu, K.; Simon, J.D. Direct chemical evidence for eumelanin pigment from the Jurassic period. *Proc. Natl. Acad. Sci. USA,* **2012**, *109*(26), 10218-10223.

[26] Carlson-Newberry, S.J. Newberry Sydne J. Emerging Technologies For Nutrition Research. National Academy Press Publishers: Washington, USA. **1997**, pp 675.

[27] Vaughan, M.R.; Quaggin, S.E. How do mesangial and endothelial cells form the glomerular tuft? *J. Am. Soc. Nephrol.,* **2008**, *19*(1), 24-33. 1

CHAPTER 3

Water of the Cephalous Spinal Fluid- The Main Source of Energy of the Central Nervous System

Arturo Solís-Herrera*, **Graciela Landín-Miranda, Ruth I. Solís-Arias, Paola E. Solís-Arias, Martha P. Solís-Arias** and **María del Carmen Arias Esparza**

Human Photosynthesis® Study Center, López Velarde 108, Centro, Aguascalientes 20000, México

Abstract: Until today it seems such an irrefutable fact that the main source of energy of the Central Nervous System is the blood vessels. However, the bloodstream itself cannot transport energy, at most only carries substances that eventually take part in endergonic reactions. We found after twelve years of continued studies about the three major causes of blindness that glucose is just a source of biomass; the role of glycan in the living organism is highly complex but only as provision of essential metabolic intermediates, and not as energy provider. Our awe-inspiring finding was that any cell in our body, independent of its function, needs water not only as a solvent or vehicle, but as a fundamental chemical energy source. The explanation of how a eukaryotic cell could dissociate or break the water at room temperature in order to get energy is also amazing: **Melanin is to the animal kingdom as chlorophyll is to the plant kingdom**. Both substances split, dissociate or break the water molecule. As an analog, we appointed it as human photosynthesis. The pharmacologic modulation of human photosynthesis or the capacity of human body to transform light into chemical energy, offers an opportunity to open new and very efficient ways to treat several diseases representing an important area of study in the epidemiological point of view.

Keywords: Blood Vessels, CSF, Energy, Hydrogen, Melanin, Oxygen, Photosynthesis.

BACKGROUND

Glucose: A Source of Energy?

Sugar or saccharides are indispensable components of all living things. The biology of saccharides (sugar chains or glycans) is studied by Glycobiology, which is defined as the study of the structure, biosynthesis, and biology of saccharides. Glycobiology is rapidly emerging as a primary field of interest of biomolecular and biomedical research worldwide. Once considered merely

* **Corresponding author Arturo Solís Herrera:** Human Photosynthesis Study Center, Av. Aguascalientes Norte 607, Pulgas Pandas Sur, Aguascalientes, México; C.P. 20138; Tel/Fax: +524492517232; E-mail: comagua2000@yahoo.com

supporting structures, the sugars have now been widely recognized to be vital components in running the complex machinery of life itself.

Sugars control and influence almost every single aspect of the cellular processes, range from cell-cell interactions, metabolism intermediates, adhesion mechanisms, growth factor signaling, blood clotting, blood types, receptor binding, regulate the activity of hormones in the blood, direct embryonic development, and serve structural roles. However, in the light of our findings, saccharides are all these but they are not a source of chemical energy.

All cells and thereby many macromolecules in nature carry a dense and complex array of covalently attached sugar chains (called oligosaccharides or glycans). In some instances, these glycans can also be free-standing entities. Since most glycans are on the outer surface of cellular and secreted macromolecules, they are in a position to modulate or mediate a wide and complex variety of events in cell-cell and cell-matrix interactions crucial to the development and function of a complex multicellular organism. They are also in a position to mediate interactions between organisms, *e.g.* host and parasite. In addition, simple, highly dynamic protein-bound glycans are abundant in the nucleus and cytoplasm, where they appear to serve as regulatory switches [1].

Neuron, as any other cell, seems to use glucose as the primary source of energy; however, it is a misconception. When blood glucose levels fall by half from a normal fasting value (to about 2-3 mM), there are signs and symptoms indicating some cognitive impairment and, at glucose levels below 1 mM, mental confusion is evident and coma may also result from prolonged glucose deprivation. Therefore, it is an apparent firm belief that glucose is the primary energy source for the adult human brain. Although our knowledge about the critical role of glucose metabolism in the maintenance of high level brain function has grown considerably in the recent years, the various factors that regulate glucose uptake and utilization in the CNS are not well-understood.

The regulation of glucose metabolism in humans is a complex process that proceeds bi-directionally, connecting the brain and peripheral tissues. Peripheral tissues, including muscle, fat, pancreas, and liver, are responsible for direct control over the transport, synthesis, storage, and metabolism of glucose [2].

Resuming glucose cannot be considered as the main energy provider of human body; instead it is an indispensable metabolic intermediate [3].

Mitochondria

This organelle is widely considered as the powerhouse of the neuronal cell or in general terms of the eukaryotic cell. Mitochondria are found in all human cells except mature erythrocytes and theoretically perform the vital task of generating adenosine triphosphate (ATP), the molecule that cell uses for the majority of its energy needs. It is widely believed that all organisms from the simplest bacteria to humans use ATP as their primary energy currency. The energy level that ATP supposedly carries is just the right amount for most biological reactions. On the other hand, nutrients, that are the source of the chemical energy for ATP, contain energy in their low-energy covalent bonds which is not very useful to do most kinds of work in the cells, however, it is thought that the cell and therefore our body is able to increase this covalent-bonds low levels of energy, simply concentrating it, after laborious extracting processes, in ATP molecule. In the words of Trefil (1992), it is amazing that "hooking and unhooking that last phosphate [on ATP] is what keeps the whole world operating."

Furthermore, ATP is considered more as an *energy-coupling agent* and *not* as a fuel. It is not a storehouse of energy set aside for some future need, mainly due to the reason that energy cannot be stored. Rather it is produced by one set of reactions and is almost immediately consumed by another. The enormous amount of activity that occurs inside each of the approximately **one hundred trillion human cells** is shown by the fact that at any instant each cell contains about ***one billion*** **ATP molecules.** This amount is sufficient for that cell's needs for only a few minutes and must be rapidly recycled. Given a hundred trillion cells in the average male, about 10^{23} or one sextillion ATP molecules normally exist in the body. For each ATP "the terminal phosphate is added and removed 3 times each minute" (Kornberg, 1989) [4].

The total human body content of ATP is only about **50 grams**, which must be constantly recycled every day.

It is very risky to think that low power containing covalent bonds of relatively unarmed substances we eat are the source of energy that allows us to build molecules of ATP. Furthermore, ATP is simply the carrier and regulation-storage unit of energy. The average daily intake of 2,500 food calories translates into a turnover of a whopping 180 kg (400 lbs) of ATP [1], however thermodynamically this is not possible. There is no way to extract more energy than the system itself has.

Daily energy consumption of heart (about 6 kg of ATP) exceeds significantly from energy consumption of other organs [5]. However, 6 kg of ATP cannot be obtained starting from 50 g of it. On the contrary, if mitochondria are widely considered as the powerhouse of the cell, what is the source of energy of mitochondria itself?

It is striking that when ATP molecule is broken down to ADP, the chemical reaction absorbs more energy than the energy it releases.

Cephalous Spinal Fluid (CSF) and the Physiology of the Choroid Plexuses (CP)

The prime function of the CP is to secrete CSF in a poorly understood way; and in man, this is achieved at a rate of ~0.4 ml/min. The total volume of CSF is around 150– 270 ml but of this, only ~25% is in the ventricles. The remainder fills the basal cisterns.

The plexus tissue has excellent blood supply as the choroidal layer of the eye; it was estimated that the rat lateral ventricle CP receives 3– 4 ml/min/g, which is 10-fold the blood supply when compared to 0.35–0.4 ml/min/g in cerebral cortex. Human eye receives already 200 ml of blood per minute, and has a weight of 7.5g. (Adler, 1998) [6]. The tissue layers with the highest blood supply in the CNS and eye curiously have similar names: Choroid plexuses in the former and choroid in the latter, which mean "grape color" by their high melanin content.

The choroid plexuses (CPs) are leaf-like highly vascular structures lying in the ventricles. The main function of choroid plexuses is the production of the cerebrospinal fluid (CSF).

Although CPs have a unique distribution of ion transporters/channels, the mechanism of CSF production is similar to the production of fluids in other epithelia and **is again theoretically based on energy released from ATP hydrolysis**, which drives unidirectional flux of ions accompanied by movement of water by osmosis.

Total volume of the CSF in humans is estimated to be ~150–270 ml, in contrast to this relatively small volume, the rate of the CSF production is relatively high, indicating the vital importance of a refreshing current CSF. It is estimated that ~600 ml of CSF is produced over 24 h in humans, which is enough to totally replace existing CSF three to four times. The total mass of these choroid plexuses in humans is about 2 g.

Therefore, the rate of secretion of CSF by choroid plexuses appears to be 0.21 ml/min/g tissue, which is many folds higher than the estimated rate of secretion for other epithelia. Such a high rate of secretion can be achieved mainly because the structure of choroid plexus is superbly adapted to the secretion:

Firstly, the plexus tissue has excellent blood supply (see above) and secondly: the capillaries of the plexuses, in contrast to the cerebral capillaries, are fenestrated in the same way as that of the choroids' capillaries of the eye and therefore provide little resistance to the movement of small molecules.

Apart from the lower level of protein in the CSF, the composition of this fluid is relatively similar to blood plasma and this similarity might indicate that the CSF could be formed by simple ultra-filtration of blood plasma. However, careful measurement of the dialysis ratio between plasma and CSF has revealed clear evidence that the **CSF is formed by active secretion** and not just by ultra-filtration of plasma.

1. The ionic composition of the CSF cannot be predicted by the Gibbs–Donan equilibrium as could make it in the case of an ultra-filtrate system; therefore, the production of CSF is not achieved through a merely ultra-filtration process.
2. The concentration of major ions in the CSF is constant, despite relatively large fluctuations in plasma concentrations for some ions.
3. CSF is slightly hypertonic, which would not be the case if it is produced by simple ultra-filtration. These facts suggest not only that the CSF is actively secreted but also that various homeostatic mechanisms control ion transport across the epithelial layer, providing a constant composition for CSF.
4. The high rate of secretion of CSF by choroid plexus requires high levels of energy. The ATP-based energy system cannot fulfill this high requirement, instead the energy that comes from the water dissociation reaction or human photosynthesis makes it affordable [7].

Human Photosynthesis or the Unexpected Capacity of Melanin to Split the Water Molecule

With the main aim to understand and unravel the physiopathology of the three main causes of blindness in humans, we dedicated twelve years of continued studies to their mechanisms of disease. In 2002, we were able to reach an astonishing conclusion: The main source of energy of the human retina is water, not ATP.

This means that any cell in our body, independent of its function, needs water not only as a solvent or vehicle, but as a main energy source. The explanation of how a eukaryotic cell could dissociate or break the water molecule at room temperature is amazing too: **Melanin is to the animal kingdom as chlorophyll is to the vegetable kingdom**. Both substances are able to split, dissociate or break the water molecule.

The unsuspected capacity of melanin to conduct at room temperature a biochemical reaction that in laboratory needs 2000 °C is awe-inspiring. As an analog, we appointed it as human photosynthesis. But in comparison, our photosynthesis is much more efficient than that of vegetables because chlorophyll only uses between 400 and 700 nm of light wavelength. In comparison, the human version of chlorophyll, that is melanin; is capable of absorbing light or more exactly energy through the whole electromagnetic spectrum.

The main source of energy in vegetables is water, because chlorophyll transforms the radiant energy of the sun (photonic) into chemical energy by means of the dissociation of water molecule. When the molecule is broken, energy is released and hydrogen is the atom that is quenching the energy. Then, energetically speaking, the most valuable is the hydrogen, not oxygen; and the explanation is quite simple: Hydrogen is an energy carrier not only in human body but also in the whole Universe, and the CNS is not an exception; the Oxygen is a by-product of the water dissociation reaction that is toxic at any concentration. Hydrogen is just an energy carrier but not an energy resource, *i.e.*, it is necessary to inject hydrogen with energy, because hydrogen itself has no energy.

In plants: $2H_2O$ (liquid) $\rightarrow 2H_{2\,(gas)} + O_{2\,(gas)}$

In Humans: $2H_2O$ (liquid) $\rightarrow 2H_{2\,(gas)} + O_{2\,(gas)} \rightarrow 2H_2O$ (liquid) $+ 4e^-$

Furthermore, evidence indicates that melanin absorbs, besides the full electromagnetic spectrum, any kind of energy, even gravitons, and the absorbed energy is used mainly in dissociating the water molecule. In addition, we learned that when human photosynthesis is turned down in an abrupt manner, **hemorrhage and edema are** ensued in any part or tissue of our body, while when it is diminished in a chronic form, then **fibrosis and mitosis** will develop in the affected area with a severity that depends on the affection in the levels of water dissociation or human photosynthesis.

The Energy Supply to the Nucleus of Eukaryotic Cell Without Mitochondria or ATP

Cell nucleus requires energy as any organelle, cell or tissue. However, the cell nucleus has neither ATP nor mitochondria, therefore, the energy necessary for each chemical reaction or processes that take place inside the nucleus are fulfilled in a different way. It is probably that the explanation is by means of energy release throughout the water dissociation or human photosynthesis.

The energy from water is released by the melanin that is contained in melanosomes located in the cytosol, and it is symmetrically distributed in all directions along the cytosol and nucleoplasm. Thus, the low-molecular-weight sugar nucleotides that act as donors for most of the biosynthetic steps can occur in both inner cell spaces due to availability of enough and constant flow of energy.

Thereafter, these are specifically transported into the lumen of the organelles; a process that requires energy.

It has become clear that certain types of glycoconjugates are synthesized and reside within the cytosol and nucleus, but the synthesis, drive and function of each substance requires energy constantly, energy is not only necessary in initial steps, instead it is required in a constant form, all the time, because from the point of view of energy, there are no differences between the beginning and the end of the thousands of chemical reactions or processes supported by this energy that is incessantly released by a cycle relatively simple and composed by two phases: water dissociation/water reformation.

When inside the melanin splitting of the water molecule occurs, molecular hydrogen and oxygen are released, and then the very important energy-carrier function of hydrogen takes place, transporting crescent spheres of energy traveling symmetrically in all directions, and when the cycle is completed, *i.e.* the reformation of the water molecule occurs; the growing spheres are formed by water however energy is released because a flow of electrons occurs. Therefore, energy with different characteristics is constantly released by melanin because the pigment requires a trillionth of second to harvesting enough energy from light to break the water molecule.

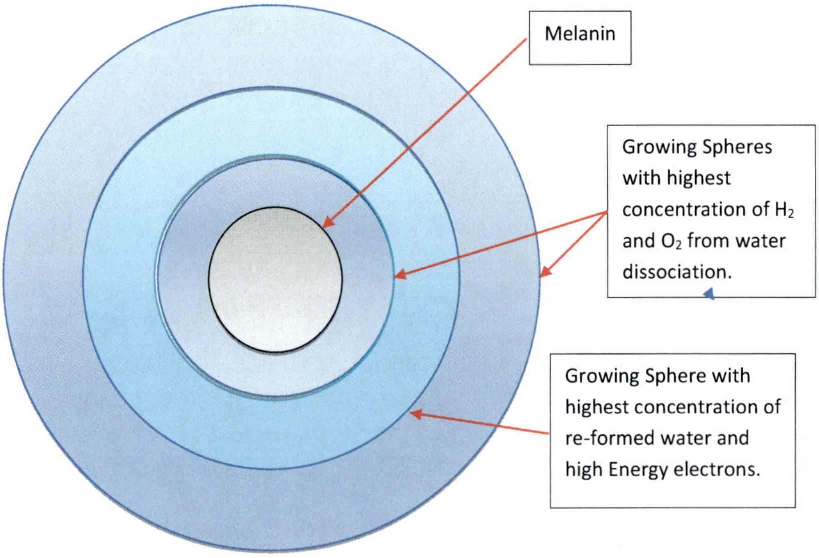

Fig. (1). Melanin constantly dissociates and re-forms water, not loose throughout, but the gas molecules tend to occupy all the space that surrounds them, so eventually deviating from the melanin, which is when they are picked up by the organelles that are found in the cytoplasm.

This question that what is the source of energy in the nucleus could be answered by the fact that melanosomes tend to form a kind of envelope around the nucleus. The energy of the human photosynthesis is released in a symmetrical form in all directions; energy emerges from melanin as in the form of spheres of alternating nature, *i.e.* gas-water or hydrogen, the energy carrier by excellence in the universe and flow of high energy electrons (Fig. **1**). This continuous flow of alternating energy spheres tends to coalesce with each other, and due to the fact that melanosomes form a type of envelope around the cell nucleus or perikaryon, the emanated energy spheres bring together to form a center of high energy (Fig. **2**).

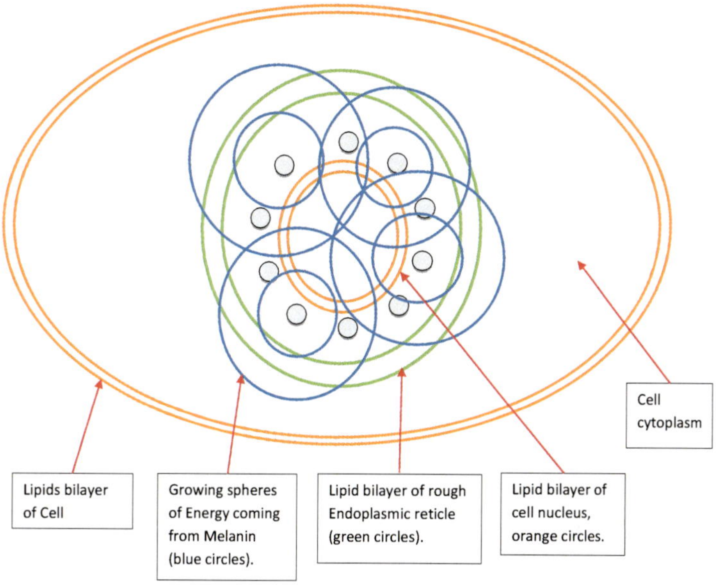

Fig. (2). Diagram representing a sagittal section of the cell, where the circles of blue color that correspond to increasing energy spheres from melanin, which is located mainly in the perinuclear space; and these growing spheres coalesce forming zones of high energy, for example to the inside of the cell nucleus, which was to be expected given that it is the cell organelle of larger size and metabolic activity, but with lack of mitochondria and ATP. On the other hand, both the cell nucleus and the melanosomes are surrounded at the same time completely by an organelle whose metabolic activity is also intense: the endoplasmic reticulum, rough, (green circles); and that provision allows the latter to capture large amount of growing energy spheres, which uses, without a doubt, for its also constant metabolic activity.

In fact any intracellular structure benefits from the energy that carries the hydrogen molecule and its powerful anti-oxidant effect, besides that for every two molecules of water that are re-formed, four high-energy electrons are generated, therefore these electrons are shared easily.

The chemical energy that comes from the melanin is capable of explaining all chemical reactions that occur in any cell of the body at any given moment, due to that any process involves exchange of energy, even prior to any change to occur is necessary in the presence of chemical energy, because the initiation of any chemical reaction is something akin to a slope upward that must be overcome so that the following phases of the process occur, is what is called the activation energy.

Any chemical process requires energy to start, during the course of the same, and for its completion, therefore energy should be present all thc timc. And it is precisely the way in which melanin releases energy, constant, incessant, shaped within relatively narrow ranges, despite the ups and downs that happen all the time in the amount of luminous energy that is present. Thus, although there is much or little light energy, the chemical energy that comes from the melanin is maintained within a surprisingly stable range. This is: when luminous energy is too strong, melanin decreases it, and when energy is low, for example in winter, melanin is able to raise it.

Only this is the possible way of explaining the extraordinarily complex biochemical processes that make up life. The basis of all cellular operation had to be something almost immutable, able to compensate for the vagaries of the environment that happen almost every minute. The intricate machinery of the cell is so complex, so intricate, so interconnected, that variations by small, that would be able to derail everything.

On the contrary, if the fundamental basis of life, which is the chemical energy, remains constant, almost unchanged, then the rest of the system has more flexibility, but the base of the pyramid may not have too many changes, as the rest of the structure would fall apart easily.

Melanin has firmness and stability necessary to do so, no doubt, and it is no coincidence that it is the most stable molecule known, as shown we are talking about 170 million years ago. It is very probable that is the most stable molecule in our body. Based on this, the understanding of the process of life, becomes easier.

The energy requirements of the cell are huge and very strict. Its characteristics do not vary more than small limits, tolerable for a self-sustaining chemical system that we call life. And that energy stability, which is essentially can not be explained on the basis of the mitochondria, as they are organelles that appear and disappear,, do not seem to have a fixed place within the organization and cell cytoplasm arrangement.

We would be talking about sources of energy that at times are here, sometimes there, sometimes many, sometimes few, and the most complicated is that at the same time, the unlikely energy function of mitochondria, would be dependent on the somewhat irregular exogenous supply of glucose and oxygen. There are too many variables to be able to cause and sustain complicated processes that lead to life.

The processes of life are very orderly, if we combine oxygen with glucose in a test tube we get not only CO_2 but also other products such as CO (carbon monoxide) in variable proportions. And in the cell, by far, the main product of metabolism is CO_2, which is the more oxidized state of carbon.

Suffice it to say that if energy source is glucose, diabetics would fly. Glucose is a wonderful molecule, but not so much to assume that it is able to provide the energy needed by its own metabolism. It would be fantastic, but nothing is further from reality.

Glucose and nutrients in general, provide the cell and organisms, with the necessary elements for biomolecules to form. This is: carbon chains of different lengths and combinations with other elements, *e.g.* nitrogen. But nothing more. Our body takes energy, defined as everything that produces a change; of water, as the plants; and the word metabolism means continuous change, therefore the requirements of chemical energy to the interior of the cell are also continuous, constant, incessant; otherwise, life is not possible.

Melanin and the Central Nervous System

The discovery of the true source of power of the structures of the nervous system, involves a new era in the diagnosis and treatment of neurological diseases. When the levels of chemical energy coming from the melanin are kept as closely as possible to what have been during all evolution, for all creation, any tissue can show an amazing resistance to the lack of circulation, hypoxia [8], and CNS is not exception.

The differences between treat cerebral vascular events with the method traditional, surgical and pharmacological stimulation of the water molecule dissociation are enormous. Surgery produces very severe sequelae both anatomical and functional (Figs. **3**, **4** and **5**); while the therapeutic administration of QIAPI 1 leaves no sequelae neither anatomic nor functional (Figs. **6** and **7**).

So far the surgery to evacuate the haematoma or extravasated blood is practiced when you consider that the chances of recovery are minimal or that there is an imminent risk to the life. However, when the intracellular chemical energy levels are adequate, tissues, whatever show a remarkable resistance to the lack of circulation or low levels of oxygen.

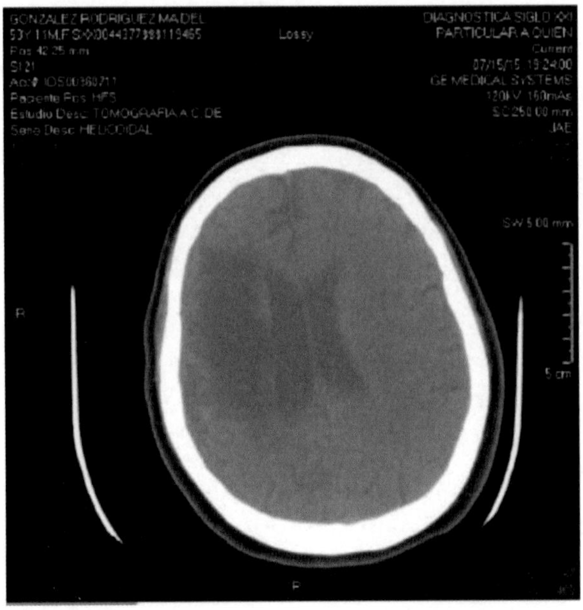

Fig. (3). CT scan of a patient with a cerebral vascular event within the territory of the middle cerebral artery. The remnant, postoperative damage is extensive, as well as the functional sequelae (spastic hemiparesis).

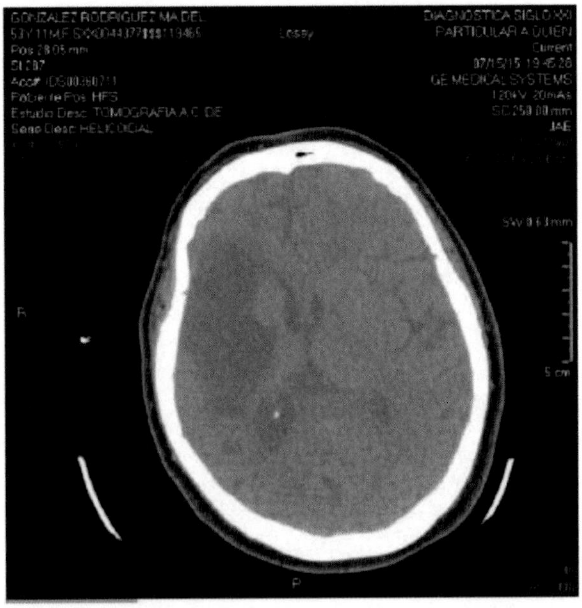

Fig. (4). Post surgical anatomical deformation is very extensive, the affected tissue shows very significant changes in the area intervened. Functional changes are disabling.

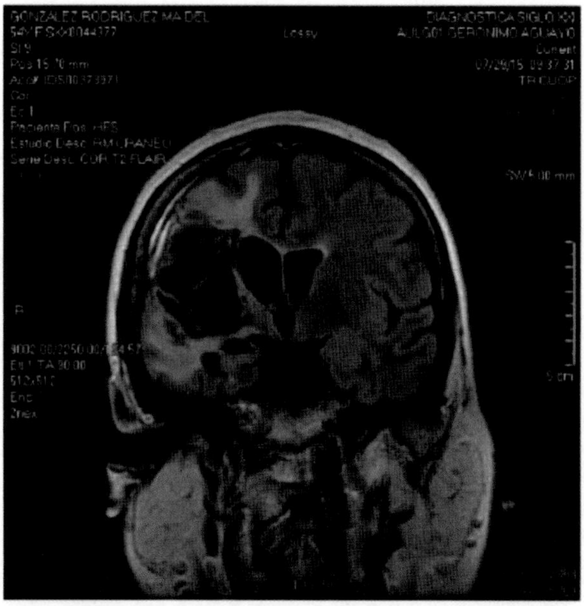

Fig. (5). Nuclear magnetic resonance study shows more clearly the enormous disruption of brain structures when cerebral vascular events are managed with surgery.

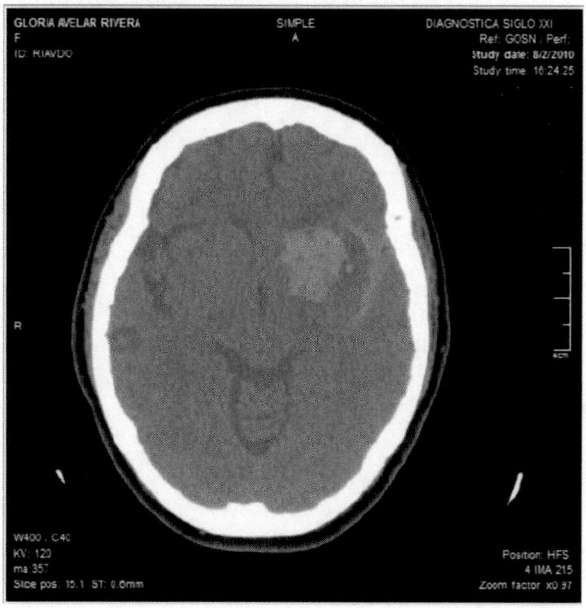

Fig. (6). On the contrary, when a cerebral vascular event, like the one shown here, is managed according to the generation and distribution of energy from the melanin, the evolution is very different, since the patients recover completely in the term of a few weeks.

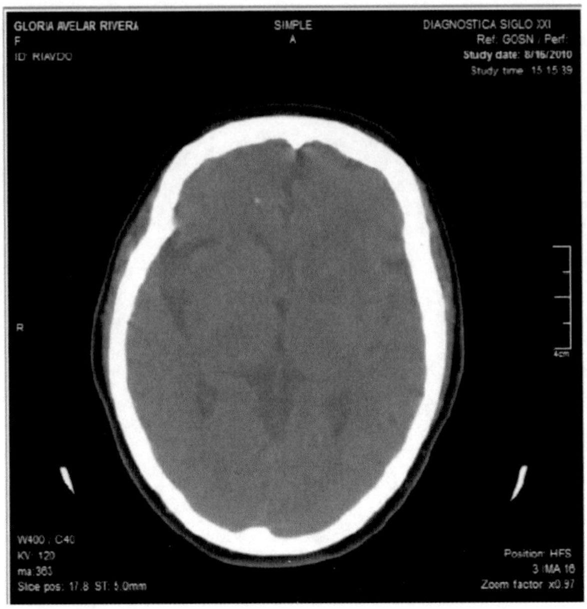

Fig. (7). The studio, 14 days after the initial event, shows a surprising recovery of the tissue affected by vascular event of the Middle cerebral artery. Anatomical features show slight changes to the neurological examination, there were no significant sensory or motor disturbances. The patient completely recovered his faculties in less than one month.

The difference in the time of recovery of patients is very notorious, as patients treated with surgery tend to have permanent sequelae, while those treated medically, through the administration of drugs that intensify the process of dissociation of water; on average, three months; already perform their normal, usual activities with minimal or no limitation [8].

In the other case, also of cerebral vascular event, in the territory of the Middle cerebral artery, computed axial tomography shows compatible findings with vascular event. The scan was made a few hours after the patient began with alterations of consciousness.

The study was evaluated by a neurosurgeon who indicated to prepare the patient immediately for surgery, but as he was halfway through another surgery, and that also of vascular event. it took 6 hours, during which the family administered you sublingually, QIAPI 1™ every hour, three drops. Six hours later, when the neurosurgeon came out of surgery, he could not believe what he was seeing, as the patient had regained consciousness, was well focused on three areas, and body mobility was surprisingly recovered. The doctor ordered a new scan that showed that swelling had decreased by 50% with only six hours of treatment. It was not possible to conduct such a study, so we did another study 14 days later (Fig. **7**).

CONCLUSION

Modern biochemistry and molecular biology confirm Newton's axiom, according to which, all Life's systems are run on the same lines of simplicity, as physical phenomena. Melanins do comply with this law, being sometimes difficult to prove due to their rather strange (for us) physical and chemical behavior.

The function of the cerebral ventricles and subarachnoid space requires rethinking, as in the light of the discovery of the intrinsic property of melanin transform light energy into chemical energy by means of the dissociation of the molecule of water, as well as the chlorophyll in plants; it is necessary to rethink the physiology of the CNS.

The current concept that the energy of the CNS is transported through the bloodstream has resulted in erroneous medical attitudes which cause severe damage to the nervous tissue. The drug-induced intensification of the generation and distribution of energy that starts in melanin shows that nervous tissue is surprisingly resistant to lack of blood supply.

The nerve cell as well as the CNS glia is the result of 4 billion years of evolution. The biochemical processes that occur inside them are mostly unknown or at least misunderstood, but have been well established by evolution, which we now understand, starts, depends on, and is governed by the generation and distribution of energy, such as the universe itself.

HUMAN AND ANIMAL RIGHTS

No humans were used in this research. All animal research procedures followed were in accordance with the standards set forth in the eighth edition of Guide for the Care and Use of Laboratory Animals published by the National Academy of Sciences, The National Academies Press, Washington, D.C.

CONFLICT OF INTEREST

The author (editor) declares no conflict of interest, financial or otherwise.

ACKNOWLEDGEMENTS

We appreciate the steadfast support of the Center for Studies of the Human Photosynthesis® to conduct this study.

REFERENCES

[1] Varki, A.; Cummings, R.; Esko, J. *Essentials of Glycobiology*; Cold Spring Harbor Laboratory Press, **1999**.

[2] Hertz, L.; Dienel, G.A. Energy Metabolism in the brain. *Gluc. Metab. Brain.,* **2002**, *51*, 1-102.

[3] Solís-Herrera, A. Beyond Mitochondria, what could be the Energy source fo the cell? *Cen. Nerv. Sys. Med. Chem.,* **2015**, *15*, 32-41.

[4] Solís-Herrera, A.; María del Carmen, A.E.; Solís-Arias, R.I.; Solís-Arias, P.E.; Solís-Arias, M.P. The unexpected capacity of melanin to dissociate the water molecule fills the gap between the life before and after ATP. *Biomed. Res.,* **2010**, *21*(2), 224-226.

[5] Chodorowski, Z.; Rybakowska, I.; Anand, J.S.; Kaletha, K. Energy metabolism in the failing heart. *Przegl. Lek.,* **2009**, *66*(6), 356-358. [Polish.].
[PMID: 19788151]

[6] Hildebrand , G.D.; Fielder, A.R. Anatomy and Physiology of the Retina. In: *Pediatric Retina*; Reynolds, J.; Olitsky, S., Eds.; Springer: Berlin Heidelberg, **2011**.

[7] Solís-Herrera, A.; María del Carmen, A.E.; Martha Patricia, S.A. Energy Production, the main role of Melanin in the Mesencephalon. *J. Appl. Med. Sci.,* **2013**, *2*(2), 11-20.

[8] Solis-Herrera, A.; Arias-Esparza, M.C.; Landin-Miranda, G.; Solis-Arias, R.I. The pharmacologic intensification of the water dissociation process, or human photosynthesis, and its effects over the recovery mechanisms in tissues affected by bloodshed of diverse etiology. *Int. J. Clin. Med.,* **2011**, *2*, 332-338.
[http://dx.doi.org/10.4236/ijcm. 2011.23058]

CHAPTER 4

The Unsuspected Intrinsic Property of Respiratory Pigments to Dissociate, Irreversible, the Water Molecule

Arturo Solís Herrera*

Human Photosynthesis® Study Center, López Velarde 108, Centro, Aguascalientes 20000, México

Abstract: According to current dogma, the oxygen that enters in the organism through exchange places with the external environment, must reach out to all cells of the body; which implies important energy expenditure, and must also do so at a speed proportional to the metabolic demands of tissues.

It is a deeply rooted dogma that molecules classified as respiratory pigments have merely a passive role as carriers of oxygen. Our finding of the inherent property of melanin to transform light energy into chemical energy through the dissociation of the water molecule, as chlorophyll in plants; will mark a before and after in the concept of respiration, with enough universal validity to be a new ground rule in Biology.

Melanin can dissociate and re-form the water molecule. Respiratory Pigments can dissociate, irreversible the water molecule.

The association between respiratory pigments and high levels of oxygen, is due to the constant dissociation of the water molecule inside the respiratory pigment, more than a passive role in the transport of oxygen from the lungs to the tissues.

Keywords: ATP, Chlorophyll, Hydrogen, Melanin, Oxygen, Photosynthesis, Respiration.

INTRODUCTION

In the animal kingdom, there are four types of O_2-binding (respiratory) pigments, and are the hemoglobin, hemocyanin, the hemerythrin; and the chlorocruorins, all with different structures but very similar functional properties. Structure of binding site vary. Prosthetic group of globin is protoheme (Fe (II)-protoporphyrin, which can bind one ligand. They have characteristic colors in their oxygenated

* **Corresponding author Arturo Solís Herrera:** Human Photosynthesis Study Center, Av. Aguascalientes Norte 607, Pulgas Pandas Sur, Aguascalientes, México; C.P. 20138; Tel/Fax: +524492517232; E-mail: comagua2000@yahoo.com

Arturo Solís Herrera (Ed.)

states [1]. All of them are considered metalloproteins. Chlorocruorin is a hemoprotein but with heme component which differs from protoheme in one substituent. The metal is iron or copper and is attached to an organic complex. In the copper protein hemocyanin and the iron protein hemerythrin. The binding site in each case contains two metal atoms. The combination with oxygen occurs in this complex; it is there where the respiratory pigments are combined and detached from molecular oxygen. There is many different hemoglobin, but all contain the same heme group; due to it is the basis of its structure; which is considered a ferrous protoporphyrin. Therefore, all the hemoglobin heme groups are identical.

Chlorocruorins and hemocyanin are always found dissolved in the blood plasma, hemerythrin occur only intracellularly; and hemoglobin are both intra and extracellular. Intracellular respiratory pigments have molecular masses under 100 kDa and only 1 to 8 O_2-binding sites per molecule. Most of extracellular blood pigments have far larger molecular masses of up to several million kDa and often more than 100 O_2-binding sites.

However, these interesting data, must be re-thinking at light of discovery that compounds with heme groups can dissociate irreversible, water molecule. Thereby, oxygen does not come from atmosphere and lungs or their equivalent, instead, comes from water dissociation inside the cell.

In the same way, the carbon dioxide coming from all the cells of the body should reach the areas of the body where it can evacuate. Some organisms have all cells enough near places of exchange with the outside, so the simple diffusion and microscopic convection currents inside the cells and tissue liquids are sufficient for driving and proper gas exchange.

Many other bodies are not constituted by molecules and fluids with the physicochemical characteristics of the blood and thereby other body fluids play an important role in the transport of gas compatible with the metabolic needs of the tissues, which to date is referred to as respiratory functions of these body fluids.

Given the enormous importance that until now was given oxygen in the production of energy, we will deepen in the characteristics of the molecule, one of the most stable of which are known.

Oxygen Transportation

An organism can be described as a self-sustaining chemical system, but far from being physically isolated from their environment, is in a state of continuous exchange with the outside. The exchanges include, inter-alia: energy, water, minerals, atoms isolated or in combination with other atoms forming molecules, ions; *etc.* Until now it was thought that obtaining oxygen and removing carbon dioxide closely relating to the use of the chemical energy contained in food.

Theoretically, to the catabolism of carbon carbohydrates, lipids and proteins into smaller molecules and thereby smaller content of chemical energy compounds, gradually, the chemical energy that was liberating, are incorporated into molecules of ATP (adenosine triphosphate), which would later be used in the various processes in which the organism or the cell requires energy.

The maximum yield of this energy could be obtained through a very complete oxidation and that would at the same time linked with an electron transport system. The chain of electrons could not function without oxygen, as it is the final acceptor of electrons. Most of cells, but not all; does not survive long without the proper functioning of a system of electron transport, which seems to explain the constant demand for oxygen.

Apparently, the two main products that we can detect in the complete oxidation of food are the carbon dioxide and water, which is called metabolic water; but in any case, part of the water content of the body. Carbon dioxide is produced continuously, but animals, unlike plants, cannot be used to synthesize other organic compounds, and if they do it should be in small proportions. After energy, the exchange of respiratory gases is the most pressing of the exchanges with the environment.

In the air, dry, to one atmosphere of pressure (760 mm Hg), the partial pressures of oxygen, nitrogen and carbon dioxide are: 159, 594; and 0.23 mm Hg respectively. When the air is saturated with water vapor, it occupies space, so the proportion of oxygen decreases and its partial pressure at 20 °C and 760 mm Hg, is 155.6 mm Hg.

When it is important to know the concentration of a gas, the quantity of gas means concentration per unit volume. The concentration of oxygen in the air, is the weight of oxygen per liter or moles ($1/100^{th}$ of the percentage of its concentration) per liter. You must always consider the temperature and pressure.

To a partial pressure, the gas concentration decreases when the temperature increases. Gases tend to dissolve in aqueous solutions. Oxygen molecules are dissolved or distributed between the water molecules, almost in the same way that the molecules of glucose or sodium chloride they do. There is gas bubbles, these would represent molecules that are not dissolved. If you put water without oxygen in contact with air with a partial pressure of 159 mm Hg, air should tend to dissolve in the water until the concentration reaches equilibrium.

If the partial pressure of oxygen in the air drops to 140 mm Hg, then the water begins to lose oxygen until again the balance is reached. This is: the partial pressure of any gas in solution, is exactly equal to the partial pressure of the same gas in gaseous phase with which the dissolution is in equilibrium.

A solution exposed to a partial pressure of a gas phase, only dissolve part of the gas, which is called absorption coefficient. For example, the absorption coefficient of oxygen in distilled water at 0° C and 760 mm Hg of pressure, is 49 ml. On the other hand, the coefficient of absorption of nitrogen and carbon dioxide, under the same conditions (0 °C and 760 mm Hg) are 24 and 1713 ml/l respectively.

The solubility of gases is greatly reduced when the temperature increases. For example, the absorption coefficient of oxygen in distilled water, at 760 mm Hg and temperatures of 0° C, 20 ° C and 40 ° C, are: 49, 31 and 23 ml/l respectively. On the other hand, the increase in the salinity decreases the solubility of the gas, *i.e.* there is an inverse correlation. The coefficients of absorption of oxygen to 0 °C, in water 0 $^{0}/_{00}$, 29 $^{0}/_{00}$, and 36 $^{0}/_{00}$, increase in salinity are: 49, 40 and 38 ml/l respectively.

This allows us to observe that the partial pressure of a gas is not sufficient data to calculate the gas dissolved in the water, temperature, pressure and salinity are variables that affect the final concentration in the water. For example, oxygen is more soluble in cold water.

Gases tend to spread from areas of higher partial pressure to areas of lower partial pressure. Therefore, it is possible that cold water has a higher concentration of oxygen, although lower oxygen tension than hot water. It is very important to know the partial pressure of the gases to understand their behavior. In any system, the equilibrium is only reached when the partial pressure is completely uniform.

Unlike nitrogen, oxygen and other atmospheric gases, carbon dioxide can react with water and form ions, bicarbonate and carbonate. Sea water can dissolve much more carbon dioxide than the distilled, before reach the equilibrium with the atmospheric gas phase.

Something similar happens with the oxygen in the blood of animals. Hemoglobin is one of respiratory pigments which can be combined with oxygen. Therefore, the molecules of oxygen combined with hemoglobin does not contribute to the partial pressure of the oxygen dissolved in the blood. This leaves only grasp it when the concentration of dissolved oxygen as such, increases to the point where the partial pressure is equal to the source of oxygen.

The oxygen concentration is much higher in the air which in water, and the fresh water is greater than the water salty at the same temperature. *i.e.* the water provides a major source of oxygen than water.

If a hypothetical animal tried to extract a liter of oxygen from your environment, 0 °C, oxygen partial pressure 159 mm Hg, the animal would have to make their surfaces respiratory 4.8 l air, 98 l of fresh water, or 125 liters of sea water.

The difference between the air and the water is even more pronounced at higher temperatures. Therefore, it is no coincidence that the highest metabolic rates have been found in animals that breathe air. In invertebrates, insects; in vertebrates: birds and mammals. The concentration of oxygen in the air, when the oxygen tension is constant, down 8% when the air temperature increases from 0 °C to 24 °C, but in fresh or salt water, descends a 40% as the water heats up 0 °C to 24 °C.

The percentage decline in the concentration of oxygen in the water with the temperature increase, is much more pronounced than in the air, and given the much lower concentrations in the water, it is conceivable that high temperatures constitute one further threat would be for aquatic species for terrestrial species.

Oxygen diffuses a few times faster 300 000 in air than in water. Carbon dioxide diffuses more slowly that the oxygen in the air, but the water diffuses about 25 times faster than oxygen. That the speed of diffusion of gases are of considerable importance for the analysis of the physiology of respiration.

The density of air, to 760 mm Hg of pressure and temperature 17 °C, is 0.0012 g/ml, while that of fresh water under the same conditions is 1 g/ml, which is some 800 times higher. On the other hand, viscosity, defined as the measure of resistance internal to the flow; it is different in different respiratory channels or capillaries. Not all areas of fluid current move at the same speed, due to the forces of cohesion between the fluid and the walls of the duct. The outer concentric layers of fluid move more slowly than the more central.

Referred to the development of resistance by friction mainly between adjacent layers of fluid moving at different speeds, viscosity. The viscosity of water is 35 times greater than the air at 40 °C and more than 100 times to 0 °C.

The higher density and viscosity of the water requires that, within certain limits; water breathing animals invest more energy to move a certain volume of the medium in question within their ducts and respiratory surfaces than land animals; which is interesting since it would mean that much of the captured oxygen would be the effort metabolic capture more oxygen; Thus, evolutionarily; it makes no sense.

1-2% of the man resting metabolism is involved in ventilation of the lungs, while perhaps more than 20% of the metabolism of the fish at rest, would go to the ventilation of the gills. The density of water is almost the same as the of the protoplasm, so water keeps the Gill filaments of aquatic animals next to a neutral suspension; which, physiologically speaking, of unusual importance because it allows the entire length of the Gill filament is in constant contact with the water.

It is a common observation that the gills of an aquatic animal hanging from a wet dough when exposed to air. On the other hand, the respiratory structures of terrestrial animals that protrude into the interior of the medium, must have enough stiffness so they sustain themselves, because otherwise, they would bend or they would deform and eventually would meet each other; with the consequent reduction of the surface exposed to breathing.

Any component or member of environment, exchanges oxygen and carbon dioxide with the neighboring parts by means of physical processes such as diffusion and convection. The speed of transfer of gas diffusion between two media depends on the difference of tension with distance. Speed drops when it reduces the tension difference and distance increases. Although diffusion can provide a very quick transfer of gas to microscopically small distances, the transfer in centimeters or decimeters is slowly. On the contrary, convection, *e.g.* in form of wind or water currents, acts independently of the pressure difference and can quickly mix air or water even at great distances.

The effect of convection and the diffusion equilibrium, can be seen in the composition of dry air, whose oxygen concentration is of 20.95% and carbon dioxide is 0.03%. The rate of carbon dioxide is which shows greater variation than the other major components of the atmosphere, but does not usually exceed the 0.06%, even in the streets of the city in the peak hours, therefore changes in the partial pressures of oxygen and carbon dioxide are due mainly to changes in barometric pressure and water vapor.

It is a fact that the tensions of oxygen and carbon dioxide show a wide range of values in the different habitats of the animals. The physical processes of diffusion and convection tend to balance the water masses with the atmosphere. The tensions of oxygen in open seas tend to be rather high, even to abyssal depths. Surface waters tend to be richer in oxygen, due to the exchange with the atmosphere and the photosynthesis of the plankton.

The intensity of light decreases with depth, and in deep water, theoretically there should be no photosynthesis or very reduced. But the reality is different, because it is an observed fact that the deep waters of the seas have a high pressure of oxygen, and is explained by the movement of the water, but we think that it is due to melanin can absorb any type of energy and dissipate it through the dissociation of the water molecule. And the enormous pressures in the depths of the sea are no exception, as it is a type of energy that melanin can absorb and dissipate it dissociating the water molecule.

All living organisms show extensive correspondence between the structure of their component molecules and active chemical processes. This is the molecular logic of the organism that starts from melanin, not from glucose.

There are many similarities at molecular level which cannot be satisfactorily explained either by functional necessity or by chance. This is particularly clear in the comparison of nucleic acid and protein sequences of different organisms as bacteria and mammals. The homology in the macromolecules is the strongest evidence for the common origin of all living organisms; in our opinion, the first spark of life comes from melanin.

It will be shown that the enormous number of different proteins can be arranged into just few hundred groups of proteins with significant homology, as proteins super-families; which apparently came into being a common ancestor during the early development of life on earth, and melanin fill requirements to be considered as common ancestor or more exactly the origin of life.

A good example are enzymes, but their catalytic properties, specificity, activity, affinities and so on, of a given enzyme vary little with the source [2]. The slight physical differences, are usually unimportant, and the enzyme remains essentially the same enzyme. Other example is the uniformity in the macromolecules carrying information, as DNA and RNA. Thereby, there is an astonishing uniformity in the bio- diversity of biochemistry.

In many aspects, the variety of living things defies the imagination. The number of species on the earth cannot be calculated, just estimated.

The Exchange Membranes

There is always a membrane that separates the internal tissues of the animal's environment. In unicellular organisms, it is the cell membrane. In a fish, the membrane would be Gill epithelial cover that separates water from the blood vessels and in humans would be the epithelium of the alveoli. The passage of molecules of gas through the membrane is by diffusion. And curiously is has failed to show any example of active transport in the animal Kingdom, except the carbonic anhydrase [1].

Gases diffuse more quickly than the ions. The diffusion of oxygen does not occur unless there is difference between both sides of the membrane oxygen tension, and the greater is the difference, happens more quickly. On the other hand, speed increases as the exchange surface is greater, so the respiratory membranes of the animals and humans form complicated models with invaginations and evaginations that increase its extension.

Likewise, the speed of diffusion varies inversely with the thickness of the membrane and the highest density of the medium in which have cross. So, you might think that the regulation of the process begins from there. For this reason, in the lung, the entrance of oxygen is air to water, *i.e.* a medium in which moves quickly to a medium in which travels 300 000 times slower. The diffusion in water is too slow as only way to provide gas transportation between the medium, the respiratory surface and the inside of the body.

Ventilation or movement convective medium (air or water) through respiratory surfaces, it can be passive or active (suction or positive pressure), but the latter implies energy expenditure and the first does not ensure the oxygen supply nor that carbon dioxide is emitted with the fast enough.

Oxygen diffuses more slowly in tissues than in the water. And given that the movement of oxygen is only by simple diffusion, then only it is efficient at very short distances. The speed of diffusion in tissues is a third of the oxygen reaches into the water. Therefore, an efficient exchange of gases can be as cells are very close to the respiratory surfaces, as in small organisms, or convective gas transport inside the body in the case of larger organisms. One of the main roles of the circulatory system is the transport of gases by convection. Oxygen can be picked up by the blood that circulates through the respiratory surfaces and transport it by convection through the body, therefore the anatomic-functional characteristics of the circulatory systems and the physical-chemical characteristics of blood are determined by the metabolic needs of the cells.

In cells, there has been a movement to the inside of the cell membrane. Referred to as cytoplasmic streaming or cyclosis. In the Protozoan Paramecium, the metabolic rate is 1 cm^3 O_2/g/h. To be elevated, thought that the simple diffusion is not enough for their metabolic needs, so the cyclosis could have a role in the transport of gases inside the cell.

Most of the body surface in mammals is little permeable to gases. Any thin membrane exposed to the environment, can serve as area of exchange of gases with the environment, especially when it is vascularized. Usually referred to as respiratory organs are the lungs in mammals, frogs; the gills of the fish; Gill papules of the starfish; but that does not mean that other parts of the body also serve classified as respiratory.

The skin of frogs and the tube feet of echinoderms are important in gas exchange. Specialized respiratory structures tend to be characterized by fine membranes in complex models, invaginated or evaginated; that it increases the surface area for gas exchange. Structures of the body evaginated and surrounded by the environment are called gills, which are invaginated in the body and contain the medium are called lungs.

Convective ventilation of respiratory surfaces, can be active or passive, one or bi-directional. The use of oxygen is the amount of gas captured by the respiratory surface of the total available.

External respiration: sponges, flat and coelenterate worms lack circulatory system, so that the bodies of these organisms are constituted in such a way that optimize the transport of gases. Sponges are perforated by many pores, and most of the cells are arranged near the outer surface of the animal, and walling throughout the pore and the channel aqueous, attracting water by flagella movement; and exits through oscula, which are large openings.

The spaces between the cells are occupied by mesenchyme, with a small number of cells (amoebocyte) separated from the exchange with the water surface, so his supply of oxygen remains in the mystery. Usually, cells with increased demand for energy, as the nerve and the muscle, are placed near the surface, either internal or external of the body. It is thought to be most gas exchange in animals that lack circulatory system through the skin or its equivalent.

Other organisms, such as the annelids; already have a quite enhanced circulatory system, which allows the cells to be farthest from the respiratory surfaces. In general terms, the respiratory features are similar on mollusks that breathe through gills, in Lungfish gastropods; in crustaceans, terrestrial and semi-terrestrial crabs in Amphipod and isopod crustaceans; in the pans of the Moluccas, in echinoderms, sea cucumbers; in fish, Teleost, elasmobranchs, eel, amphibians, *etc*.

We can say that they are a respiratory exchange surface, air, and water. The word oxygen is mentioned dozens of times, for the supposed importance in energy metabolism. But if for a moment, we think that oxygen is not as important and the excretion of carbon dioxide is thousands of times more, then the descriptions and the approach of the various authors would not change much. The handling of the gas is very important, but not in the sense of the oxygen uptake but that carbon dioxide is expelled continuously.

Our discovery of the intrinsic property of melanin transform light energy into chemical energy through the water molecule dissociation, means nothing more or nothing less that the cells can produce its own oxygen. Therefore, it is not necessary to bring it from the outside, what we need is expel carbon dioxide, which, the anatomical infrastructure is the same.

The same anatomy works for the gases to pass from the cell to the air or to the water, and in the case of the carbon dioxide the important thing is that they leave the cell or the organism. Even the need for the media bathing the respiratory surfaces to be replaced more or less continuously is due to saturation with carbon dioxide, not to the low oxygen voltage.

To review the features of the oxygen, we realize that life would be almost impossible. For example, the oxygen in the water moves 300 000 times more slowly than in the air. Therefore, when oxygen (theoretically) passes from the pulmonary alveoli into the blood, makes it even more slowly, which would not be sufficient to provide or to blood same not far to the tissues of the oxygen necessary for the so brought and led controlled combustion of glucose. Perhaps the idea that our body "burns" the glucose for energy, emerged with Lavoisier ("plants produce oxygen"). And then no one had tried to compare.

Since then looks like an obsession trying to explain how the oxygen in the atmosphere manages to reach the inside of the body and even more to each cell. It is so rampant the attempt that researchers undergo high situations physiologically impossible. Based on that could not be found any mechanism that increases oxygen uptake in any organism, the explanation of the alleged greater need for oxygen during exercise remedied with increased respiratory and heart rate; but they both also work in the case of carbon dioxide. And so really, because the body concerned more carbon dioxide than oxygen, otherwise the bodies would have accessories mechanisms to capture greater amount of oxygen during ventilation, as in normal situations only absorbs one-third of the oxygen, a toxic gas. Nature has this capacity, indeed.

However, there is a mechanism widely disseminated to increase greatly carbon dioxide expelling and is carbonic anhydrase or carbonate dehydratase. This enzyme is found not only in all interchange surfaces instead is almost ubiquitous, there are 6 subtypes described, but its basic mechanism of action is the same one.

Accelerates the transformation of carbon dioxide into bicarbonate, and does it in both directions. Even is the enzyme with high pKa which is known: one million molecules per second. So, important is the management of dioxide of carbon for nature. Death from asphyxia is not the hypoxia or anoxia, is due to the accumulation of carbon dioxide.

Hemoglobin

The basic molecular unit of the hemoglobin consists of a heme group (Fig. **1**) attached to a protein part or globin. All the hemoglobin heme groups are identical. The physical and chemical properties of the hemoglobin changes in different species, and in any species, you can find several different hemoglobin.

Heme Group

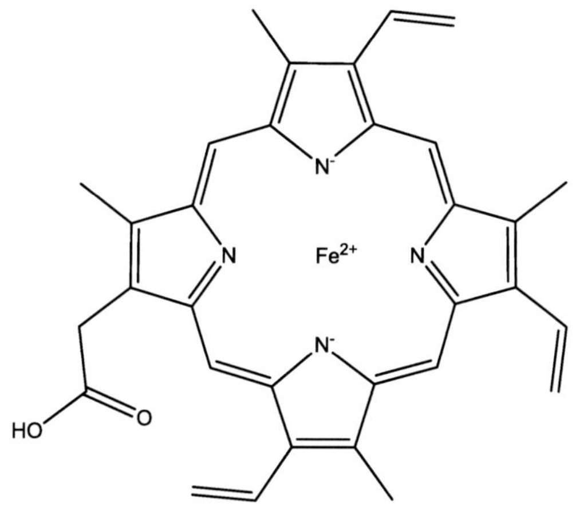

Chemical Formula: $C_{32}H_{28}FeN_4O_2$
Exact Mass: 556.16
Molecular Weight: 556.44
m/z: 556.16 (100.0%), 557.16 (37.3%), 558.16 (7.6%), 554.16 (6.4%), 555.16 (2.3%), 557.15 (1.5%)
Elemental Analysis: C, 69.07; H, 5.07; Fe, 10.04; N, 10.07; O, 5.75

Fig. (1). Heme consists of a porphyrin ring that holds a central iron. Porphyrin is the name of purple -colored stone. Besides hemoglobin, heme is also found in many cytochromes, and comprises various classes of enzymes and electron carriers.

The combination of the heme with the oxygen group is not an oxidation of iron to ferric Heme State. If so, it could not carry oxygen and is what happens with the hemoglobin goal. The heme, as other porphyrin, shows a characteristic absorption spectrum, and can be, although slightly, influenced by the nature of the globin part. The unit of the hemoglobin molecule is a group heme and globin associated unit, which often are unique to form large molecules. In vertebrates are usually units of 4 molecules.

The similarity of the heme group with chlorophyll (Fig. **2**) is notable. On the other hand, the hemoglobin are most widely spread respiratory pigments. They are the only respiratory pigments of vertebrates and their blood contains hemoglobin. The hemoglobin of muscles are called myoglobin and vertebrate always appear as a single molecule units. They tend to be concentrated in different muscles active, such as the heart; to who gives the characteristic reddish color.

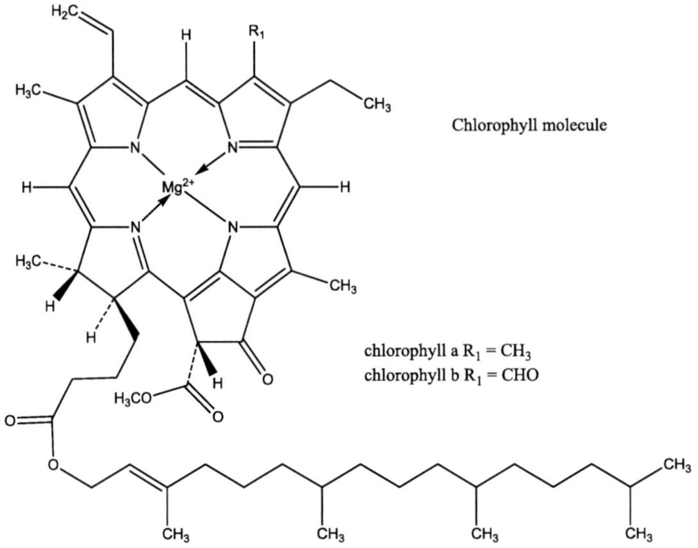

Chemical Formula: $C_{55}H_{71}MgN_4O_5^{2+}$
Exact Mass: 891.53
Molecular Weight: 892.48
m/z: 445.76 (100.0%), 446.26 (73.2%), 446.76 (41.1%), 447.26 (11.2%), 447.27 (4.2%), 447.76 (2.9%), 446.26 (1.5%), 447.77 (1.2%)
Elemental Analysis: C, 74.02; H, 8.02; Mg, 2.72; N, 6.28; O, 8.96

Fig. (2). Chlorophyll is essentially two parts: a substituted porphyrin ring and a long carbon chain so-named phytol. The porphyrin ring is a chelating ligand with the four nitrogen atoms, binding strongly to a coordinated metal atom in a square planar arrangement. Heme consists of a porphyrin similar to that in chlorophyll but with an iron ion in the center. Vitamin B12 contains a cobalt ion at the center of the porphyrin ring.

Usually the insects tend to lack of respiratory pigments, but the amount of melanin in the exoskeleton is remarkably high. The cytochromes of electron transport consist of groups of proteins with similar to heme iron-porphyrins. Given his enormous resemblance with chlorophyll, it is conceivable that the hemoglobin are also able to dissociate the water molecule.

Hemocyanin

After the hemoglobin, hemocyanin are the most widespread respiratory pigments. Very little to understand its structure; they contain copper, usually with high molecular weights. Hemocyanin are blue when oxygenated and colorless when they are deoxygenated.

Chlorocruorins

The distribution of the remaining respiratory pigments, the chlorocruorins and the hemerythrin, is considerably more limited. The chlorocruorins are chemically; many similarities with the hemoglobin, as the basic unit consists of a Porphyrin Iron. The chlorocruorins are always found in plasma in solution, and are polymerized in large molecules. They are greenish coloration. Sabelido *Protamilla* presents hemoglobin in the muscles but chlorocruorin in blood.

Hemerythrin

The hemerythrin contain iron, but not in a porphyritic complex. They are purplish red when oxygenated and colorless when they are deoxygenated. Its iron concentration is 2 to 3 times higher than in the hemoglobin of vertebrates.

Oxygen and Respiratory Pigments

The combination of the pigment with each group sensor of oxygen from a respiratory pigment is stoichiometric, which means that only a molecule of oxygen is transmitted with a hemoglobin heme group. But it is more likely that molecular oxygen comes from the hemoglobin itself than atmosphere, given the enormous difficulties above mentioned for the displacement of oxygen in different media. On the other hand, to dissociate the hemoglobin molecule of water in an irreversible manner, as well as chlorophyll; molecular hydrogen and oxygen diatomic; occur being the major product hydrogen, given that the hauler par excellence of energy in the universe is an integer; and oxygen, a very stable molecule; it is toxic at any concentration. Although cells and organisms optimize it and therefore used it in some biochemical processes.

Clearly has more advantage considering that hemoglobin also transform light energy into chemical energy, which regard it as just a somewhat passive oxygen carrier. It really does not make much sense hemoglobin to passively capture oxygen; well anyway, it would require the presence of chemical energy to do it and undo it; which it is not possible from the orthodox point of view given that red blood cells do not possess mitochondria. If we consider that the hemoglobin produces energy, then it would be understandable how erythrocytes fill its energy needs, which as any cell, are constant and varied, as the cells use the energy in many way, even to keep the shape.

The affinity of the hemoglobin to change automatically and exactly between oxygen and carbon dioxide, depending on the different stages of blood circulation, is less complicated to explain if we assume that the increase of CO_2 in the blood decreases the process of dissociation of water that occurs in hemoglobin, thereby the partial pressure of oxygen decreases more due to descent in the dissociation of water rather than the alleged recruitment of the oxygen by the tissues.

And when venous blood reach the pulmonary alveoli, the blood CO_2 is moved very quickly to the alveoli due to the great activity of the enzyme carbonic anhydrase and physicochemical properties of CO_2 in relation to water; therefore we descend the concentration of carbon dioxide in the blood, so the ability of hemoglobin to dissociate the water molecule recovers, and molecular oxygen levels coming from the water molecule dissociation returning to soar.

Human alveolar oxygen tension is about 100 mm Hg, when the atmosphere is about 160 mm Hg, and when venous blood reaches the lung the PO_2 is 40 mm Hg, and blood coming out of the lung is 95 mm Hg. But this contradicts the observation that only a third of the oxygen in the alveoli passes into the blood, and on the other hand, if the oxygen is so important then organisms would have mechanisms that accelerate the entrance of oxygen to the blood, and to date has not found any. And to further complicate things, the speed with which moves the oxygen in the air is about 300 000 times faster than in water, and in blood is even slower [2]. And the solubility of oxygen in the blood is about 40 ml/l or less.

Besides the lungs never fully deflate with an exhalation, therefore the inspired air mixes with this residual air, lowering the partial pressure of oxygen within the alveoli. This results in a lower concentration of oxygen in the lungs than is found in the air outside the body.

Even for nature had been complicated solve many challenges and obstacles if it is to have the desire of transporter oxygen from the atmosphere to all and each one of the cells of the body through a circulatory system with abundant respiratory pigments. Oxygen and carbon dioxide move independently of each other.

On the other hand, if we look at the data in relation to carbon dioxide, we find the following: the PCO_2 in the atmosphere is 0.23 mm Hg. And its solubility in blood is around 1700 ml/l; venous blood reaching the lung has a PCO_2 of about 45 mm Hg, while within the alveolus the PCO_2 is 40 mm Hg. The blood that leaves the lung presents a PCO_2 of 40 mmHg.

Transport of CO_2 from the tissues to the blood and then to the lungs is crucial, because the CO_2 levels greater than 45 mm Hg down turn the process of generation and distribution of energy due its wide effects on water and melanin itself, which results in increased cellular dysfunctions and eventually cell death. Despite the continuous production of carbon dioxide by all cells of the body, levels of PCO_2 in blood ranging from 40 and 45 mm Hg. But it's a process strictly regulated by 4 billion years of evolution it is not something that happens randomly, and different mechanisms are responsible. For example, in addition to CO_2 transported in hemoglobin, in places other than oxygen; it also dissolves in important quantities in the plasma by the action of the enzyme carbonic anhydrase that is also widely present in all cells, or at least one of its 6 subtypes.

And here could be a question, the energy certainly need the carbonic anhydrase enzyme to carry out its functions in the plasma or cells, where does?

And if the answer is that it comes from the dissociation of water that irreversibly happens in hemoglobin, in a similar way to the chlorophyll, compound with which presents many similarities; and that is conveyed by the diatomic hydrogen, then the explanation is more consistent.

The dissociation of the water in the hemoglobin molecule is as rapid and constant as in chlorophyll, so the production of H_2 and O_2 never ceases, and gases diffuse from the inside of the erythrocyte, where they occur; and eventually reach the blood plasma. Since hydrogen is the smallest atom, they pass easily through the various inner membranes containing erythrocytes. And as you move, without combined with water; hydrogen is captured in one way or another to be used in various chemical reactions by its powerful antioxidant effect both their energy charge, as it is able to reduce to the oxygen itself forming water.

Eventually, the H_2 comes to the plasma, where the function is the same, energy carrier and anti-oxidation, which explains not only the function of carbonic anhydrase, but also other biochemical processes that are known to occur in plasma, but that date remain poorly understood. It is somewhat similar to the biochemical changes that occur in the SAP of plants, whose circulation through it and the biochemical processes that occur in the meantime, had neither a satisfactory explanation; but melanin which contains the trunk of the plants, which also transforms light energy into chemical energy, will help in their understanding. Recall that any chemical reaction involves energy interchange.

It is difficult to accept that CO_2, despite so many advantages that it has compared with oxygen, which diffuse by simple diffusion, only can decrease in 5 mm Hg, while the oxygen increase from 40 to 95 mm Hg. An important confounding factor is that the carbon dioxide is produced continuously in all cells of the body, but still, the difference between the alveolar and arterial blood flowing from the lung seem minimal.

Venous Oxygen Tension

The different tensions of oxygen in the blood as it passes through the various tissues of the body reflect the dynamics of the organism. So far they were interpreted as an effect of the factors that affect the supply and oxygen utilization rates. But once the cell is capable of producing its own energy through the water molecule dissociation, we now understand that the oxygen supply as well as its use is not as we imagined it.

Firstly, the cell does not "burn" glucose for energy, therefore does not require oxygen from the outside. Which uses can be obtained more easily by the dissociation of the water molecule, which through the Odyssey that so far is intended to. The real supply of the chemical energy required for cell, and used in many ways; It is through the H_2 resulting also from the dissociation of water. Molecular hydrogen is the real motor of life, only glucose is source of carbon chains with which the body implements or in their case replaces the various structures that make up the body, for example: skin, nails, hair, bone, muscle, nervous tissue, intracellular organelles, enzymes; *etc.* Looking at it that way, the anatomical and functional layout of the respiratory surfaces and circulatory system is more towards management of carbon dioxide than oxygen.

The need for cells to expel carbon dioxide is huge, constant, and incessant; as occurs on an ongoing basis by the cellular metabolism. In the same way, the source of chemical energy that is required to promote each and every one of the intra-cell biochemical reactions, estimated in thousands per minute; It also has to be constant, relentless, unchanging; and the melanin does, because it transforms light energy visible and invisible constantly, day and night.

Above mentioned that the combination and the oxygen uptake are stoichiometric, that means that only one molecule of oxygen will join a hemoglobin heme group at the same time. Although the human blood contains about 5.4×10^{20} per 100 cm^3 heme groups, and given the many obstacles that the oxygen would have to overcome to be able to saturate all of them on their way by the lung. It seems almost impossible to happen on a daily basis. If we grant that hemoglobin heme groups are actually separating the water molecule and constantly generate H_2 and O_2, then it is more physiologically, more logical to think that to be dissociating constantly, there will always be the presence of oxygen in the groups we have; rather than think that heme groups were the lucky capture and combine with oxygen and the oxygen molecules that joined the heme are the same all the way in the human body, and more prowess even if we take into account that an average human body possesses an estimated 95 000 km of capillaries.

It is said that blood carries about 20 cm^3 of blood per each 100 cm^3. The amount of oxygen carrying plasma is small compared with the hemoglobin which transports. The amount of oxygen dissolved in plasma, solution, increases linearly with oxygen tension, but the amount of "fixed" oxygen as oxyhemoglobin shows a different behavior, as it is a curve of type sigmoidal which is affected by the amount of light, and more than 8 decades, it still said that the oxygen exchange that takes place between the atmosphere and the blood cannot be explained adequately only by diffusion. In other words, there is something else.

In experiments conducted in atmospheres with high levels of CO_2, for example 6%; they showed no significant increase in blood oxygen tension, so it is considered that elevated CO_2 levels are not a stimulus for the absorption of oxygen from the lungs.

So the mystery continues: what is that which raises nearly three times the alveolar oxygen tension when it passes into the blood? Even when rising levels of oxygen in the inspired air, the relative difference between blood and alveolar oxygen tension remains the same. On the other hand, with tensions as low as 7 or 8% (artificial) atmospheric oxygen, hemoglobin still shows a 96% oxygen saturation.

Therefore, it seems to not depend on inspired air, which supports our proposal that most of the oxygen combined with hemoglobin comes from the dissociation of the water molecule, pair part mainly hemoglobin and to a lesser extent by the melanin. Which would explain that despite the enormous difficulties of oxygen for air entering the blood, what happens is the opposite; as the oxygen that cells produced by melanin, hemoglobin; or both, tends to move to the alveolus, passively. Not so carbon dioxide as physic-chemical properties of gas as well as the omnipresence of the enzyme carbonic anhydrase, accelerate significantly his expulsion from the blood into the alveoli.

The unsuspected capacity of the hemoglobin of irreversibly to dissociate the water molecule, explains the persistence of the double bands of the spectrum of oxyhemoglobin, which persist for a long time even after the animal breathing has stopped. It is very interesting Haldane´s observation *"In bright weather the room was also partially darkened during tritation"*.

It is well known that low temperatures reduce the transformation of light energy into chemical energy in plants; so it is not strange that the fall in body temperature causes an also marked fall of blood oxygen tension.

Oxygen Tension and Exercise

The blood carries oxygen in simple solution in the plasma; as well as in the form of oxyhemoglobin. As a solution, the amount of oxygen is small and increases linearly with oxygen tension; but the amount of oxygen fixed as oxyhemoglobin seem to show a powerful sigmoid relationship with partial alveolar oxygen tension. I could tell that the effect of the respiratory pigment is considerably greater than the simple dissolution of oxygen. Evolutionarily speaking makes more sense respiratory pigments to dissociate the water molecule, so it always appear elevated levels of its oxygen to her around, think that nature developed only a pigment that carries oxygen. The evolution is considerably faster and cheaper with a respiratory pigment able to transform light energy into chemical energy.

A tension of approx. 20 mm Hg oxygen predominates in the blood that leaves the muscles in different situations of exercise. To increase the intensity of the workout and theoretically increases demand for oxygen by the tissues, in particular the muscles; venous oxygen tension and the coefficient of oxygen are surprisingly stable.

To date, this is explained by the increased flow blood to tissues, which can be increased in forms system, but to some extent, beyond which the system tends to collapse.

How do you explain that if you increase the distribution of oxygen to meet the increased demand of oxygen, does not grow the transfer of oxygen for each unit of blood volume? Theoretically, the explanation is based on increasing the flow of blood circulation in muscles, allowing the venous pressure remains stable.

During strenuous exercise, the oxygenation of the blood coming from the muscles can drop almost to zero, which is interpreted, wrongly as a complete deoxygenation of the blood. But in the light of the discovery of the ability of the human body to transform light energy into chemical energy by means of melanin and of respiratory pigments, through the dissociation of water, such as chlorophyll molecule; in plants, the explanation could be different: strenuous exercise, produces an exhaustion so intense, that the transformation of chemical energy into light energy falter significantly, so it is no coincidence that one of the complications of strenuous exercise is sudden death.

When the blood coming from the muscles drops below 10 mm Hg, appears mitochondrial dysfunction, as it is estimated that it requires a voltage of 0.2 at 2 mm Hg in the mitochondria directly to carry out its function (theoretical) generate energy.

Fortunately, the observed values range between 16 and 20 mm Hg in tension of oxygen in the blood that comes out of the muscles. Which explains why blood is mixed with other tissues that seem to not use so much oxygen to function properly. Mixed blood oxygen tension is a valuable indicator, but does not provide information about the download in the tissues.

The curve of dissociation of human myoglobin is a hyperbolic curve very similar to the hemoglobin. As the molecules are going "oxygenating", the process seems to be provided and then the curve becomes sigmoid. But we think that increasing the availability of chemical energy carried by the resulting hydrogen dissociation of water by the respiratory pigment hemoglobin and myoglobin, then the activation energy is achieved more easily and quickly. The above confirmed the result of experiments that demonstrates that the oxygenation of the heme groups, does not alter the other.

Often an increment of CO_2 tension is associated with a decrease in oxygen tension, but from our point of view, is associated in reality when the dissociation of water by the respiratory pigments is affected in some way, *e.g.* cold, anesthetic agents, heavy metals, *etc.*; available chemical energy decreases and they the system begin to have faults, like for example the activity of the enzyme carbonic anhydrase, and on the other hand, the levels of oxygen, that actually come from inside the red blood cells; they also reflect a decrease in the rate of normal dissociation of water, which is usually in the range of the Pico and nanoseconds.

During the average time of permanence of the erythrocytes in the capillaries, ranging between 0.25 and 2.0 seconds, the transfer of CO_2 is several orders of magnitude greater than that of oxygen, and this in all organisms. Nature just insists on important things. As a point of comparison, pointed out that sea water dissolves about 0.5 Vol % to 159 mm Hg pressure and one atmosphere of pressure.

Hemocyanin and the Transport of Oxygen in Marine Animals

Hemocyanin has been studied in numerous species and the resulting model seems strange in base to the glucose as energy source but it is not if the metabolism is based on melanin. Hemocyanin, despite its relative high oxygen affinity, and although the species studied lived in well aerated waters; hemocyanin was far from saturated in the gills of the species studied.

For example, hemocyanin of Espinosa lobster (*Panulirus interruptus*) becomes virtually saturated at a tension of oxygen of only 25 mm Hg. However, in seawater well aerated, blood from the gills of this species was not saturated, then your oxygen tension was not beyond the 7 mm Hg, and hemocyanin was only 54% of the oxygen that could capture.

Studies in other species revealed that usually post-gill blood presents a saturation between 50% and 70%, which was interpreted as that of the decapod Gill surfaces may present a unique and powerful barrier to the diffusion of oxygen between blood and water.

The data seem to suggest that the respiratory pigments act only in certain situations difficult to determine, but if we assume that the respiratory pigments are able to dissociate the water molecule, the situation is much less complex and more accessible.

Tests conducted by the administration of carbon monoxide are misleading because small doses stimulate the water molecule dissociation and high doses blocked it.

CONCLUSION

Long the dogma that the oxygen present in the blood of the bodies comes from the atmosphere, as we will continue without understanding the mysteries of life. The obstacles presented by the transport of oxygen from the outside bodies inside, are formidable and it is unacceptable to think that nature failed to resolve them over billions of years of evolution, because it would seem that the structures that make up the respiratory surfaces of animals were made to obstruct rather than facilitate the transport of oxygen.

There are so many imponderables about the tensions of oxygen in the air, fresh water and salt water, temperature, atmospheric pressure and amount of light, *etc.*; it seems impossible that life could arise around so much variability.

Life, on the contrary; it starts with firm steps, and which are repeated with an amazing constancy; we refer to the transformation of light energy into chemical energy. Melanin has such capacity, as it absorbs any kind of energy and dissipates it through dissociation and subsequent re-form of the water molecule. Since it is able to absorb radiation within the range of visible and invisible light, it works during the day and night. And to be the more stable substance that is known [3], 170 million years showed, explains which produces chemical energy in the intensity and duration necessary to make the strict order which means life, could be assembled throughout eons of years of evolution.

The form and function of living organisms is determined completely by the generation and distribution of energy, as well as the entire universe. Hence the presence of patterns that are repeated over and over again in different organisms. Beyond the role of melanin during embryonic development to date, has not deserved of some lines, and they are usually limited to mentioning that melanocytes, which give the skin color, are derived from the neural crest [4].

The discovery that the human organism is able to transform light into energy chemical through the dissociation of water, as in the plants, is expected to produce a dramatic turn in many areas of knowledge, and among them will be the embryology, because while it is known with certainty that the umbilical cord arterial blood oxygen partial saturation percentage is 50% less that the arterial blood of the mother, still insisting that energy is obtained by the fetus by oxidizing glucose.

Melanin teaches us to give food, the body gets the different carbon chains with which the body builds 99% of biomolecules, but our body takes energy, defined as everything that produces a change of light, through the water; as the plants. And in the fetus is the same, from the blood of the mother takes the building blocks for the new being, but energy takes the amniotic fluid that is 99% water. And we can see it when the fetus has plenty of water or low water, because both are bad signs, since the amount of water is accurate.

HUMAN AND ANIMAL RIGHTS

No humans were used in this research. All animal research procedures followed were in accordance with the standards set forth in the eighth edition of Guide for the Care and Use of Laboratory Animals published by the National Academy of Sciences, The National Academies Press, Washington, D.C.

CONFLICT OF INTEREST

The author (editor) declares no conflict of interest, financial or otherwise.

ACKNOWLEDGEMENT

Human Photosynthesis® Study Center, S.C. gives the financial support for this work.

REFERENCES

[1] Masoro, E.J.; Siegel, P.D. *Acid-Base Regulation: Its Physiology and Pathophysiology*; Saunders: Philadelphia, **1971**.

[2] Roughton, F.J.W. Transport of Oxygen and Carbon Dioxide. *Handbook of Physiology, Section 3; Respiration. Vol. 1*; Feen, W.O.; Rahn, H., Eds.; American Physiological Society: Washington, D.C., **1964**.

[3] Glass, K.; Ito, S.; Wilby, P.R.; Sota, T.; Nakamura, A.; Russell, B.; Vinther, J.; Dutta, S.; Summons, R.; Briggs, D.E.G.; Wakamatsu, K.; Simon, J.D. Direct chemical evidence for eumelanin pigment from the Jurassic period. *Proc Natl Acad Sci USA,* **2012**, *109*(26), 10218-10223.

[4] Sadler, T.W. *Langman's Medical Embriology,* 13th ed; Wolters Kluwer Health: Philadelphia, **2015**.

CHAPTER 5

The Source of Energy of the Nucleus in Eukaryotic Cell is Molecular Hydrogen and High Energy Electrons

Arturo Solís Herrera, Arias Esparza, María del Carmen*, **Solís Arias Ruth I.**, **Solís Arias Paola E.** and **Solís Arias Martha P.**

* *Human Photosynthesis® Study Center, Aguascalientes, México*

Abstract: Until today it is a widely disseminated belief that the main source of energy of eukaryotic cell is ATP. Those means that the daily intake of food is transformed into phosphate´s compounds of high energy as ATP and GTP. Supposedly this metabolic pathway is enough for 95% of our energy´s needs. However, our work about the three main causes of blindness in Mexico, taught us an astonishing fact: 99% of the energy of human retina comes from water. Furthermore, the explanation is amazing: human body has a photo-system composed by light-melanin-water (in order of abundance in the universe) with the extraordinary capacity of uses water as source of electrons.

Keywords: Cell Nucleus, Energy, Eukaryotic cell, Hydrogen, Melanin, Oxygen, Photosynthesis.

INTRODUCTION

The nucleus is the hallmark of eukaryotic cells; the very term eukaryotic means having a "true nucleus". The nucleus is enveloped by a pair of membranes. The inner membrane is stabilized by a meshwork of intermediate filament proteins called **laminins**. The nucleus (from Latin: Kernel) sometimes referred to as the control center, contains most of the cell´s genetic material. The genes within these chromosomes are the cell's nuclear genome. The function of the nucleus is to maintain the integrity of these genes and to control the activities of the cell by regulating gene expression, the nucleus is therefore the control center of the eukaryotic cell, and every activity requires energy. The nucleus contains the chromosomes of the cell. Each chromosome consists of a single molecule of DNA complexes with an equal mass of proteins. Collectively, the DNA of the nucleus

* **Corresponding author Arturo Solís Herrera:** Human Photosynthesis Study Center, Av. Aguascalientes Norte 607, Pulgas Pandas Sur, Aguascalientes, México; C.P. 20138; Tel/Fax: +52 4492517232; E-mail: comagua2000@yahoo.com

with its associated proteins is called **chromatin**. Most of the protein consists of multiple copies of 5 kinds of **histones**. These are basic proteins, bristling with positively charged arginine and lysine residues. (Both Arg and Lys have a free amino group on their r-group, which attracts protons (H^+) giving them a positive charge). Just the choice of amino acids you would make to bind tightly to the negatively-charged phosphate groups of DNA.

Chromatin also contains small amounts of a wide variety of **non-histone proteins**. Most of these are transcription factors (*e.g.*, the steroid receptors) and their association with the DNA is more transient. The binding of histones to DNA does not depend on nucleotide sequences in the DNA but does depend critically on the amino acid sequence of the histone. Histones are some of the most conserved molecules during evolution. Histone H4 in the calf differs from H4 in the pea plant at only 2 amino acids residues in the chain of 102. It is now clear that histones are a dynamic component of chromatin and not simply inert DNA-packing material

The main structures making up the nucleus are the nuclear envelope, a double membrane that encloses the entire organelle enclosing a lumen that is continuous with that of the endoplasmic reticulum and separates its contents from the cellular cytoplasm, and the nuclear lamina, a meshwork within the nucleus that adds mechanical support, as the cytoskeleton supports the cell as a whole. The usual location of melanin (Fig. **1**) is in the perinuclear space. Because the nuclear membrane is impermeable to most molecules, nuclear pores are required to allow movement of molecules across the envelope. These pores cross both membranes, providing a channel that allows free movement of small molecules and ions but not ATP. The movement of larger molecules such as proteins is carefully controlled, and requires active transport regulated by carrier proteins; recall that active transport means energy requirements in any or other way. Nuclear transport is crucial to the cell function; therefore, movement through the pores is required for both gene expression and chromosomal maintenance. Transport through the nuclear pore complexes is **active**; that is, it requires energy as many different carrier molecules each specialized to transport a cargo that docking molecules in the NPC.

All proteins are synthesized in the cytosol and those needed by the nucleus, must be imported into it through the NPCs. Probably each of these proteins has a characteristic sequence of amino acids — called a **nuclear localization sequence** (NLS) — that targets it for entry.

Fig. (1). Photomicrograph of melanin, the human chlorophyll 40X.

They include:

- all the **histones,** compounds rich in basic amino acids, needed to make the nucleosomes
- all the **ribosomal proteins** needed for the assembly of ribosomes. All the RNAs of the cell nucleus are organized into complex ribonucleoprotein (RNP) particles. The most part of nuclear RNAs are degrade in the nucleus (energy expenditure), only about 5% of them reach the cytoplasm in the form of ribosomes.
- all the **transcription factors** (*e.g.*, the steroid receptors) needed to turn genes on (and off), also including polymerases. With only 10^{-9} to 10^{-12} errors per nucleotide per generation, the reliability of DNA replication is very high [1].
- All the **splicing factors** needed to process pre-mRNA into mature mRNA molecules; that is, to cut out intron regions and splice the exon regions. Approximately 10% of the total RNA of the cell is present in the nucleus, the remaining 90% is in the cytoplasm. Precursors an intermediates of RNA maturation show a very high turnover rate, that means energy expenditure. The mRNAs are bound to various proteins in the cytoplasm to form RNP particles, the informosomes. There are free informosomes with non-translatable mRNA [2].

It is not at random that the nucleus is the largest cellular organelle in animals; however, it had not mitochondria; therefore: Where´s nucleus energy comes from? The nuclear envelope completely encloses the nucleus and separates the cell's genetic material from the surrounding cytoplasm, serving as a barrier to prevent macromolecules from diffusing freely between the nucleon-plasm and the cytoplasm.

Nuclear pores, which provide aqueous channels through the envelope, are composed of multiple proteins, collectively referred to as nucleoporins. The pores are about 12500 Daltons in molecular weight and consist of 100 proteins (in vertebrates). The pores are 100 nm (100 microns is the width average of human hair) in total diameter; however, the gap through which molecules freely diffuse is only about 9 nm wide, due to the presence of regulatory systems within the center of the pore. This size allows the free passage of small water-soluble molecules while preventing larger molecules, such as nucleic acids and larger proteins, from inappropriately entering or exiting the nucleus. These large molecules must be actively transported into the nucleus instead. The nucleus of a typical mammalian cell will have about 3000 to 4000 pores throughout its envelope, each of which contains a donut-shaped, eightfold-symmetric ring-shaped structure at a position where the inner and outer membranes fuse. Attached to the ring is a structure called the *nuclear basket* that extends into the nucleon-plasm, and a series of filamentous extensions that reach it into the cytoplasm. Both structures serve to mediate binding to nuclear transport proteins. The nuclear envelope is perforated by thousands of nuclear pore complexes (NPCs) that control the passage of molecules in and out of the nucleus.

Aside ATP cannot be stored; hence its consumption closely follows its synthesis; most of the ATP synthesized in the mitochondria will be used for cellular processes in the cytoplasm; and the nucleus energetic needs? The nucleus is the largest organelle; therefore, its metabolic energetic requirements must be significant. A system that is far from equilibrium contains Gibbs' free energy, and is capable of doing work. Living cells maintain the ratio of ATP to ADP at a point ten orders of magnitude from equilibrium, with ATP concentrations a thousand-fold higher than the concentration of ADP. This displacement from equilibrium means (theoretically) that the hydrolysis of ATP in the cell releases a large amount of energy. The human body, which on average contains 8.8 oz (250 grams) of ATP, turns over its own weight in ATP each day.

It is remarkable the absence of ATP and mitochondria from the nuclea of the cells. Any unstable system of potentially reactive molecules could potentially serve as a way of storing free energy, if the cell or the nucleus maintained their concentration far from the equilibrium point of the reaction. However the nucleus needs a significant expenditure of energy in order to achieve this maintenance beside metabolism and many cellular processes, including biosynthetic reactions, motility, and cell division.

The presence of melanin (Fig. **1**) in the perinuclear space starts to make sense on the basis of their unexpected ability to transform the visible and invisible light into chemical energy.

Light/Melanin/Water, the Human Photo-system

The energy released by this hitherto unknown photo-system explains the striking finding of the absence of mitochondria and ATP in the nucleus of eukaryotic cells.

The plant´s photo system is composed by: Light/water/chlorophyll and the human´s photo-system is integrated by order of abundance in the Universe by: Light/Melanin/Water. The main function of both is the dissociation of the water molecule, however our photo system has the capacity to make the processes in both ways, water splitting at first and then could bring back the water molecule again from Hydrogen and Oxygen diatomic, then, depending of variables as reactant concentrations, humidity, temperature, pressure, amount of light; the process is reversible although it is not symmetrical in time; however, in vegetables it´s only in one sense: the water dissociation (Fig. **2**).

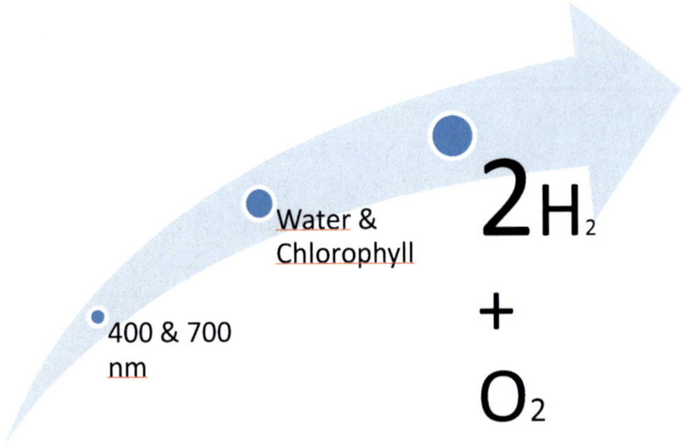

Fig. (2). The Photo system of the vegetables. Notice that the process has just one way. The dissociation of water takes place in an irreversible manner.

The efficiency of photosystem consisting of light/melanin/water in order of abundance in the universe, is thousands times more than the photosystem of light/water/chlorophyll. Chlorophyll absorbs only the ends of the visible spectrum and with that energy dissociates water irreversibly, by contrast, melanin absorbs all of the electromagnetic spectrum, allowing you to more quickly capture the energy necessary to dissociate the water molecule, but it is also capable of re - shape it, *i.e.* liquid transforms it into gaseous hydrogen and gaseous oxygen molecular in both cases, but also has the amazing ability to return it to its original liquid state (Fig. **3**).

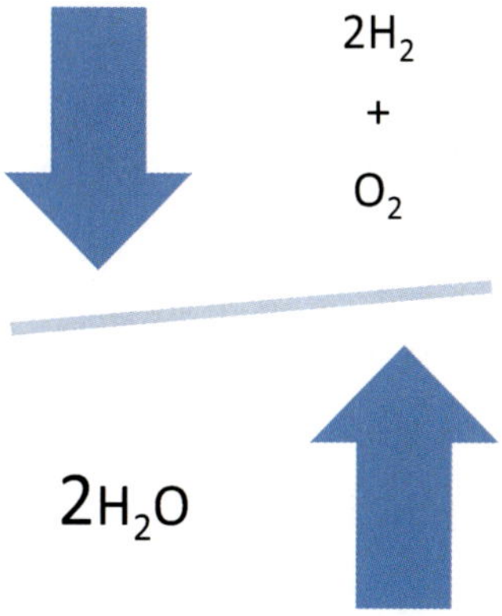

Fig. (3). The Human´s photo system. The processes are fully reversible.

The reversibility of the Human or Melanin-Photo system can be explained by mean of the next diagram (Fig. **4**):

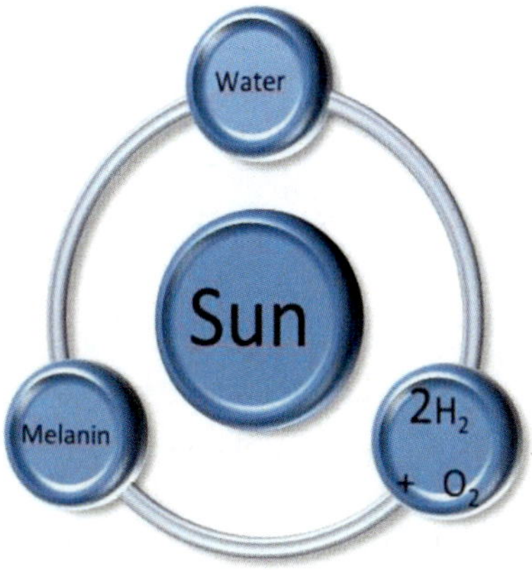

Fig. (4). When the spin of the circle turned to the right, then we will have molecular Hydrogen and molecular Oxygen, however, when its turned to the left then the process bring back water again. Both processes are driven by variables as reactant concentrations, temperature, light; pressure, humidity; other ions; *etc*. The processes are continuous, night and day, due to melanin absorbs visible and invisible light; then the process enduring decades.

Therefore, the sense of the reaction in humans is an all-time changing process in one or another way. The processes of dissociation and back bonding are not symmetrical in time. When the human photo system released diatomic Hydrogen, then the cell will have the kind of energy that Hydrogen could carried on, which is a type of energy with no mass, like photonic energy are. Otherwise, when the photo system brings back water, other form of energy will be liberated in form of an orderly flow of electrons which has the typical characteristics of any physiological action potential which rapidly rises and fall, following a defined trajectory, which means less energy and more mass. However, both kinds of energy are susceptible of being useful in one or another way for the cell and/or the nucleus. Therefore, beside two types of energy (with more or less mass), the cell and the nucleus could get water or Hydrogen, the best antioxidant.

Anatomically, the melanosomes that contain the melanin trends to surround the nucleus (Fig. **5**) like a spherical envelope [2].

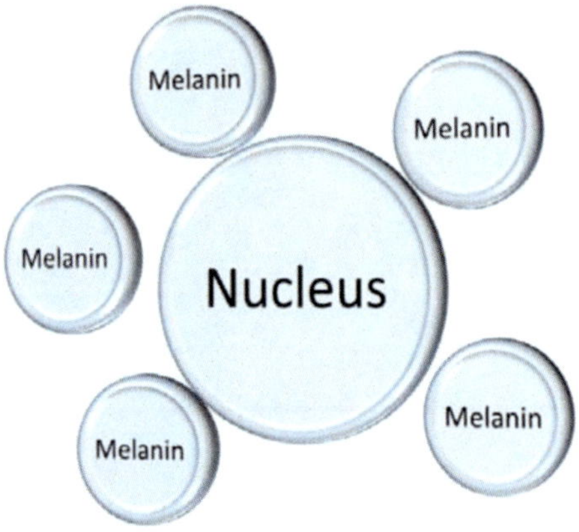

Fig. (5). The cell nucleus with the envelope of melanin. This is a repetitive pattern observed in eukaryotic cells all.

The arrangement of melanin inside the cell (Figs. **6** and **7**) is not at random manner or even by chance; instead it is a highly effective form to concentrate significant amounts of the energy released in the center of the theoretical sphere; mainly because the replacement of energy is through the concentric circles that were forming like happened on the surface of the lake when a stone is troughed on it; however into the cell the diffusion of the energy it's in a concentric spherical manner, being the zone with the highest energy those where the circles are coalescing (Fig. **6**), however the portion of energy that corresponded to cytoplasm is not affected. In other words, the nucleus will receiving higher levels of energy, essential for their control functions without take away energy that will go to the cytoplasm, which one could comprise the functions of the rest of organelles, leaving intact their functions.

The Melanin Photo-System characteristically work night and day, because melanin can absorb the whole electromagnetic spectrum; from radio waves to gamma rays [3]. Therefore, the water dissociation/re-association takes place in an endless form, 24 hour at day. Remember that an electromagnetic radiation is like a coiled string whose needs and therefore has more energy as the size of the wavelength is smallest; imagine it like when we are trying to compress a coiled string.

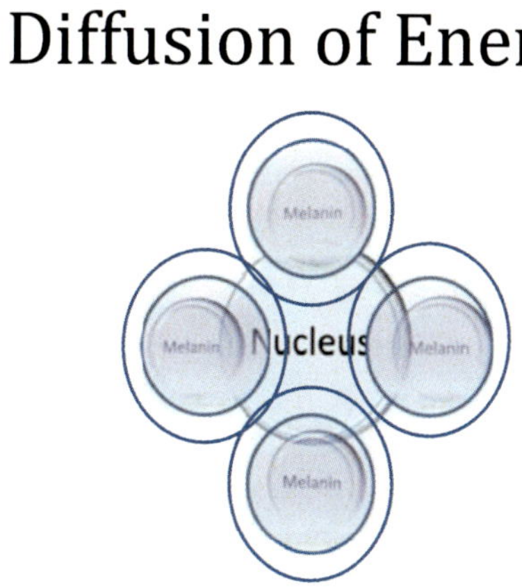

Fig. (6). The area where the diffusion´s circles are coalescing, is the zone of highest energy. The energy needs of the largest organelle, is covered in this way.

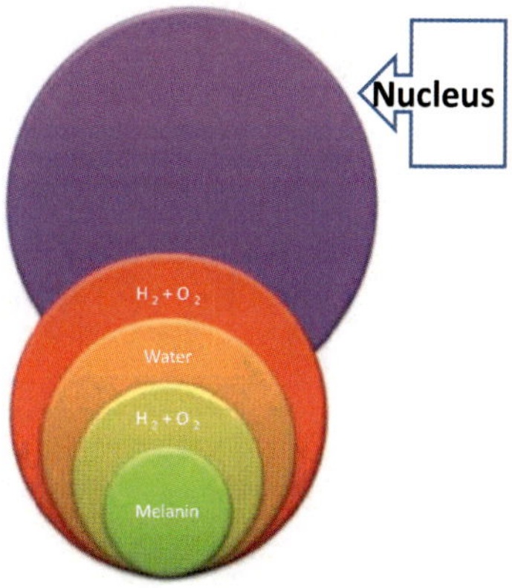

Fig. (7). In this example, the circles of diffusion (spheres) of energy will occur in an alternate form arising from melanin and goes into the nucleus, bombarded it with energy in a constant way. The nucleus is the main zone where Energy is addressed; however, the cell´s cytoplasm is bathed too by these circles/spheres of concentric diffusion.

The way in which melanin transforms light energy into chemical energy is extraordinary, as it produces energy both during dissociation of the water molecule, which is transported by the molecular hydrogen (H_2), the carrier of energy par excellence in the entire universe. 4 electrons of high energy (Fig. **8**) which have a cloud probabilistic farthest from the nucleus of the atom in question in comparison with low-energy electrons are generated as it also when re-form the water molecule, since for every two molecules of water that are re-formed. High energy electrons are easily exchanged between atoms and even between molecules.

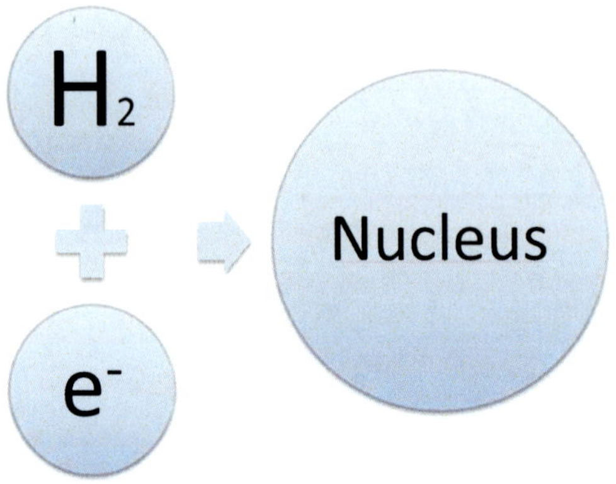

Fig. (8). Melanin Photo System produces two different types of energy. However, the nucleus cell can use both in spite their differences, in example the energy carried by Hydrogen does not have mass, and the second one (high energy electrons) has mass and different charge.

The capacity of our Photo System to split the water molecule begin to decrease at 26 years old, around 10% each decade and after fifties goes into free fall. However, many environmental risk factors exist that act in detrimental form like pesticides, herbicides; alcohol, iron´s supplements; age, sex; chemical compounds with high volume of apparent distribution; life style; diabetes, female hormones, *etc*.

In general terms, when Human Photo Synthesis is turned down in abruptly manner, there will be edema and hemorrhage, and when is turned down in a chronic form then tissue fibrosis and/or mitosis will ensue.

Astonishing, melanin Photo-system needs a trillion of second to split or dissociate the first one water molecule.

CONCLUSION

Classification of living organisms in the animal kingdom, usually presents no problem, except in specific cases. Classification of flagellate species to "Phyto-flagellates" or "zooflagellates" based on the presence or absence of photosynthetic capacity has long been considered questionable.

Paramecium, belonging to the Ciliophoran, assume such a special position amongst the eukaryotes due to of the fundamental characteristics of their nucleic acid and protein sequences, thereby they cannot logically be classified as either animals or plants. A proof that Mother Nature does not take care of human classifications.

The expected quantitative differences between low molecular-weight substances that are substrates or intermediates of cell metabolism give important information about the physiological state of cells but lead to no phylogenetic conclusions.

The complete genetic information of the organism is encoded in the order of bases, and whit it, also is the whole spectrum of genetically determined variation within and between individuals. A highly complex genetic code in route from DNA *via* RNA to proteins, so 64 possible triplet codons define only 20 amino acids. Lately, results of molecular biology research have shown genetic alteration to be far more frequent than was previously assumed. The genome appears as a dynamic system whose structure changes from one cell generation to the next, not only through single base substitution but also by rearrangement, transposition, amplification and the deletion of longer sequences. We must keep in mind that any change, any biochemical reaction requires energy of activation that comes from water dissociation, in the form of molecular hydrogen (H_2) and high energy electrons.

The cellular nucleus, as any cellular organelle, requires energy to be used in multiple ways, since any chemical or biochemical reaction involves exchange of energy. It is interesting that supramolecular structure of chromatin does not always have the well-known Watson and Crick conformation of right-handed double helix; the left-handed form is also quite often found in chromosomes. Longer sequences with purines on one strand and pyrimidines on the opposite strand can also form triple helices [4]. To date there are excellent works published about the important functions of the nucleus, but researchers seem to have not noticed the fact that the cell nucleus does not contain neither mitochondria nor ATP.

The mitochondria of the eukaryotes possess their own system of protein synthesis, that means energy expenditure. The mitochondrial genome diverges markedly from chromosomal-cytoplasmic system. Many characters involved here (mtDNA) are indistinguishable in all eukaryotes, thereby considered as homologues; however, there is a large variation in the molecular structure. Sometimes mtDNA is circular DNA double helix with a length of 15-20 or even 42 kb. Other species have larger, linear molecules, meanwhile in trypanosomes a complicated network of linked circles exists [5].

During evolution, there has apparently been a transfer of DNA between the mitochondria and the nucleus. Sequences of mtDNA-like have been detected in the nuclei of many kinds of eukaryotes [6]. Chromosomal DNA shows lower rate of evolution than mtDNA. It is interesting that mitochondria lack a DNA repair system. The mtDNA molecule shows the greatest economy of space in that there are no introns and no non-coding sequences between the genes. Cells containing more than one type of mtDNA are referred to as heteroplastic. The mtDNA of Drosophila yakuba contains the same genes as vertebrate mtDNA, although in a rather different order [7].

The ribosomes of the mitochondria are, no exception; smaller than prokaryotic ribosomes [8]. The protein of mitochondrial ribosomes form a distinct group, due to they are synthesized in the cytoplasm and only later introduced into the mitochondria [9]. Enzymes and protein factors associated with the DNA polymerases [10], such those required for unwinding the DNA and stabilizing the single strands, for the synthesis of RNA primers, and nucleases and ligases; *etc*., all them have the common requirement that is available chemical energy (from melanin).

Until today it is thought that as consistent order, so homogeneous that is observed in living beings, seems to start in the genes, but the discovery of unsuspected intrinsic property of melanin make visible and invisible light into chemical energy, through the dissociation of the molecule of water, such as chlorophyll in plants; modifies the dogma, as the order seen in living beings starts from the energy source that explains, finally; the source gives life, and not from macromolecules that carrying information.

The needs of power at the cell nucleus are intense, constant, and incessant; and each one of them without doubt, requires chemical energy. Due to not possess mitochondria, one might think that ATP can come from the cytoplasm and enter the nucleus, but it would mean more spending power besides that each molecule of ATP required to reconstitute, at least; every 20 seconds, so just outside of the core transport to the interior of the same would represent a formidable challenge that would make life almost impossible.

The study of the biochemical processes that occur inside the nucleus and the cell, were experiencing a significant change regarding the concept that our body can take energy from water, like plants, spread enough to form a critical mass of researchers that guide its work in this regard.

HUMAN AND ANIMAL RIGHTS

No humans were used in this research. All animal research procedures followed were in accordance with the standards set forth in the eighth edition of Guide for the Care and Use of Laboratory Animals published by the National Academy of Sciences, The National Academies Press, Washington, D.C.

CONFLICT OF INTEREST

Authors declare that there is no competing interest.

ACKNOWLEDGEMENT

We appreciate the support of the Human Photosynthesis® Study Center.

REFERENCES

[1] Kunkel, T.A.; Bebenek, K. Recent studies of the fidelity of DNA synthesis. *Biochim. Biophys. Acta,* **1988**, *951*(1), 1-15.
[http://dx.doi.org/10.1016/0167-4781(88)90020-6] [PMID: 2847793]

[2] Hill, H.Z. The function of melanin or six blind people examine an elephant. *BioEssays,* **1992**, *14*(1), 49-56.
[http://dx.doi.org/10.1002/bies.950140111] [PMID: 1546980]

[3] Solis Herrera, A. The unexpected capacity of melanin to dissociate the water molecules fills the gap between the life before and after ATP. *Biomed. Res.,* **2010**, *21*(2), 224-227.

[4] Lancillotti, F.; Lopez, M.C.; Arias, P.; Alonso, C. Z-DNA in transcriptionally active chromosomes. *Proc. Natl. Acad. Sci. USA,* **1987**, *84*(6), 1560-1564.
[http://dx.doi.org/10.1073/pnas.84.6.1560] [PMID: 3470742]

[5] Desjardins, P.; Morais, R. Sequence and gene organization of the chicken mitochondrial genome. A novel gene order in higher vertebrates. *J. Mol. Biol.,* **1990**, *212*(4), 599-634.
[http://dx.doi.org/10.1016/0022-2836(90)90225-B] [PMID: 2329578]

[6] Corral, M.; Baffet, G.; Kitzis, A.; Paris, B.; Tichonicky, L.; Kruh, J.; Guguen-Guillouzo, C.; Defer, N. DNA sequences homologous to mitochondrial genes in nuclei from normal rat tissues and from rat hepatoma cells. *Biochem. Biophys. Res. Commun.,* **1989**, *162*(1), 258-264.
[http://dx.doi.org/10.1016/0006-291X(89)91990-6] [PMID: 2751651]

[7] Fleming, J.E.; Walton, J.K.; Dubitsky, R.; Bensch, K.G. Aging results in an unusual expression of Drosophila heat shock proteins. *Proc. Natl. Acad. Sci. USA,* **1988**, *85*(11), 4099-4103.
[http://dx.doi.org/10.1073/pnas.85.11.4099] [PMID: 3131774]

[8] Brimacombe, R.; Stiege, W. Structure and function of ribosomal RNA. *Biochem. J.,* **1985**, *229*(1), 1-17.
[http://dx.doi.org/10.1042/bj2290001] [PMID: 3899100]

[9] Shannon, M.F.; Duke, E.J. Comparison of mitochondrial and cytoplasmic ribosomal proteins in Drosophila. *Compo Biochem. Physiol. Pt. B,* **1985**, *81*, 683-686.
[http://dx.doi.org/10.1016/0305-0491(85)90386-4]

[10] Campbell, J.L. Eukaryotic DNA replication. *Annu. Rev. Biochem.,* **1986**, *55*, 733-771.
[http://dx.doi.org/10.1146/annurev.bi.55.070186.003505] [PMID: 3017196]

CHAPTER 6

Mental Illness, Alterations in Ventricle Size, and Human Photosynthesis®

Arturo Solís Herrera, María del Carmen Arias Esparza, Ruth Isabel Solís Arias, Paola Eugenia Solís Arias and **Martha Patricia Solís Arias**

Human Photosynthesis Study Center, Lopez Velarde 108, Centro. Aguascalientes 2000, México

Abstract: This paper addresses the actual role of CSF in the physiology and ethiopathogeny of mental disorders as schizophrenia, schizoaffective disorder and bipolar disorder. Several decades ago the observation that the enlargement of ventricular size was associated with a clinical diagnosis of schizoaffective illness was reported [1] by first time.

Changes in the volume of the cerebral ventricles and therefore the content of the liquid cephalous spinal far beyond the daily physiological variations, are already known in mental disorders, but have only been to date more a curious object of in-depth studies for various reasons. Usually it pays more attention to alterations of the nervous tissue in which volumetric changes are observed in the ventricles. The finding that the melanin molecule has the amazing ability to transform light energy into chemical energy through the dissociation of the molecule of water, such as chlorophyll in plants; the old dogma that the source of energy of the Central nervous system (CNS) is through blood vessels break into a thousand pieces. We now know that the real source of energy of the CNS is the visible and invisible light and the transducer per excellence is melanin, and whose perfect substrate is water.

It is, therefore, that the CNS of all mammals has a repeating pattern invariably to be coated inside and out by the CSF that is 99% water, which ensures that the energy substrate source, water for all practical purposes, is constantly in contact with neurons. The brain caesuras allow water getting as much as possible to the most recondite places of the brain. Thereby the source of energy of CNS are the ventricles and subarachnoid space, therefore, the changes in volumetric characteristics of this anatomic regions must be interpreted as significant alterations in generation and distribution of chemical energy.

Keywords: CSF, Melanin, Mental Illness, Ventricle size, Volume.

* **Corresponding author Arturo Solís Herrera:** Human Photosynthesis Study Center, Av. Aguascalientes Norte 607, Pulgas Pandas Sur, Aguascalientes, México; C.P. 20138; Tel/Fax: +524492517232; E-mail: comagua2000@yahoo.com

BACKGROUND

Brain, ventricular size and the ratio of brain size to total intracranial volume have been a topic of interest since the advent of capacity to image the brain [1]. Walter Dandy introduced pneumo- encephalography (PNG) in 1919, replacing cerebral spinal fluid with air, which made it possible to study the contours and major morphological changes in the brain [2]. Moore *et al.* in 1935 suggested that similar abnormalities founded in patients with dementia and organic psychoses could be demonstrated in patients with the so-called functional psychoses [3]. In 71 patients with schizophrenia and 46 patients with manic depressive psychosis, these authors found evidence of cortical atrophy in much of patients with schizophrenia. In manic depressive psychosis, these investigators stated "The encephalograms in this group showed no consistent picture that would characterize manic-depressive psychosis".

In 101 cases of schizophrenia of which 73 has a diagnosis of definite or probable dementia as well, Hauge [4] reported evidence of abnormal PEGs in 58%, usually ventricular dilatation or increased subarachnoid space suggestive of cortical atrophy.

Computerized axial tomography greatly enhanced the capacity to visualize the outlines of the brain and ventricular system and identify significant structural abnormalities, although volumetric calculations were compromised by issues of radiolucent properties of CSF and slice thickness. It is interesting that studies in larger numbers of subjects appeared to less often find significant differences compared to studies with fewer subjects [5].

Magnetic resonance imaging (MRI) led to a very significant increase of neuro-anatomical studies of brain structure in mental illness (Fig. **1**). The Internet Brain Volumetric Database attempts to archive this extensive volumetric data [6].

Despite over a hundred years of research on the topic, specifics remain obscure, with studies using meta-analyses most often supporting evidence of increased ventricular volume and selected decreases (cortex and hippocampus) in brain volume [7].

Computed tomography (CT) studies have provided evidence of enlarged ventricles and enlarged cortical sulci in large proportions of persons with schizophrenia (Fig. **1**). These findings suggest a decrease in brain volume. In those patients who have these findings, the enlargement of ventricles and sulci is evident at the onset of illness, and does not progress over the course of the illness. Enlargement of the ventricles may be more common in males. Ventricular enlargement correlates with larger numbers of hospitalizations. A complimentary finding has been reported by some, but not all, authors. They have found that skull volume is reduced by 3-5% in persons with schizophrenia. Since brain growth drives skull growth, these findings suggest that the process causing schizophrenia takes place prior to the completion of brain growth (approximately age 18). Ventriculomegaly consistent data are of undoubted theoretical interest in accounting for the etiology of schizophrenia [8].

Table 1. Normal group volumes.

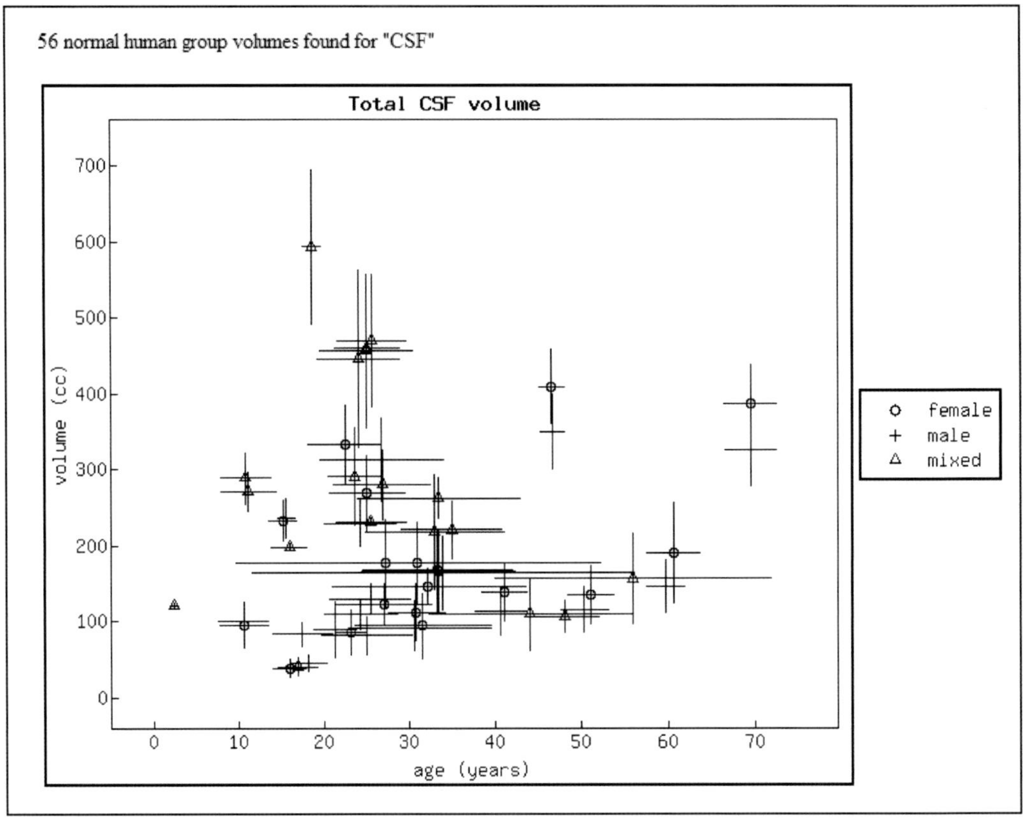

Table **1** Non-parametric distribution of 56 normal humans group volumes found for “CSF”. Source: The Internet Brain Volume Database. The issue of adequate controls for the study of ventriculomegaly in schizophrenia has been explored initially by Andresen *et al.* (1982), who proposed that healthy volunteers, a group with presumably smaller ventricles than medical controls [9]. There is a difference in VBR between schizophrenics and controls which would seem to be an indisputable characteristic of schizophrenia. The raised questions are: Which are the mechanisms responsible for the size of the ventricles? What are the molecular or tissue mechanisms implicated in the ventricular enlargement of schizophrenia?

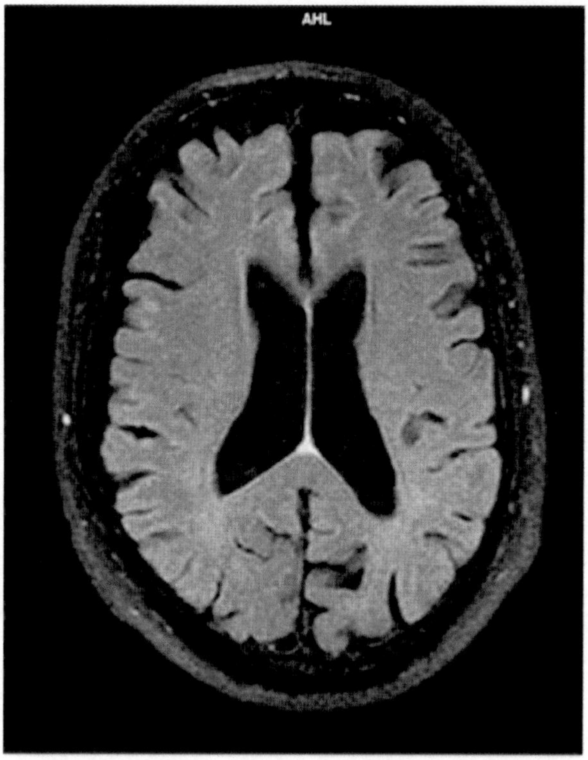

Fig. (1). Example of a patient with an enlarged both ventricles and the subarachnoid space.

The so-frequent lumbar puncture, where opening pressure of the cerebrospinal fluid are measured and liquid analyzed, has proven not be useful in the study of schizophrenia. Epidemiological studies had led to the identification of three risk factors: genetic characteristics, gestational and birth complications, and winter birth [10]. Geographically, a more benign course of illness is observed in developing, as compared with developed (cold, less sunlight), countries [11].

The dopamine hypothesis provides the most enduring explanation of the biochemical pathophysiology of schizophrenia, but there is no decisive evidence for it. The most compelling evidence comes from reports of increased levels of dopamine or its metabolite, homovanillic acid (4-Hydroxy-3-methoxyphenylacetic acid), a phenol found in human urine, and increased numbers of dopamine receptors in postmortem tissue from patients with schizophrenia. The discovery of additional dopamine receptors (D3, D4, and D5), and knowledge of interaction between meso-cortical and mesolimbic dopaminergic systems, has led to a modification of the dopamine hypothesis [12].

Pneumoencephalographic and computed tomographic studies have consistently reported abnormalities in the ventricular system in patients with schizophrenia [13]. Magnetic resonance imaging (MRI) studies have replicated the earlier observations of ventricular-system abnormalities and have demonstrated structural abnormalities in the temporal lobe, included a decreased volume of the amygdala and hippocampus [14]. Structural and metabolic abnormalities have been observed in the prefrontal white matter in patients with schizophrenia [15].

In spite that ventricular abnormality is the most robust structural finding in schizophrenia [16], the cause usually is attributed to developmental delay in childhood. The significant deficits in cortical grey matter, temporal grey matter and reduced thalamic volume are less consistent evidence of schizophrenia than the enlargement of the ventricular volumes. However, the preferred explanations cited in the literature are developmental delays and/ or atrophy of the cortical brain tissue. The water content of the CSF is seemed as static, irrelevant, or so stable factor that simply fill the empty space that the retraction of the brain tissue due to the cortical atrophy left behind. Perhaps it is not the best explanation.

In accordance with our findings in human retina at first that the main source of energy of the eukaryotic cell is water through the unsuspected intrinsic property of melanin to split the water molecule [17], the role of the water content of CSF is highly dynamic. If we assume that glucose is just a very important biomass source but water dissociation is the real source of energy, then the ventricles volume is much more important than we are thinking because their volume are determined by the extent of the water dissociation/re-association ratio.

The explanation is probably the following:

Until today, single molecule able to transform light into chemical energy by means of the water molecule dissociation, was chlorophyll, which was initially proposed by Lavoisier and Prietsley, in the second half of the 17th century. Current biology in its entirety and therefore medical practice including research, evolve around this idea.

However, our discovery of an unsuspected inherent capacity of melanin make visible and invisible light into chemical energy through the dissociation of the molecule of water, like as chlorophyll in plants it involves a substantive reorganization of biology and medicine.

For example, the sacred role of the glucose as energy source, par excellence, of eukaryote cell; now breaks into a thousand pieces. Glucose is still a fundamental molecule, but as a source of carbon chains, which our body modified in various ways to form 99% of biomolecules that make up our body. But glucose is not able to provide at the same time, mass and energy; *i.e.* glucose cannot provide the energy that is required for their own metabolism. Therefore, the energy that our body requires continuously, gets it light, separating the water molecule, like plants.

The basic reaction that occurs incessantly, night and day into every single melanin molecule, in presence of visible and invisible light and water, is the so-called Solis-Herrera cycle [18] and can be described in this way:

$$\mathbf{2H_2O\ (liquid) \leftrightarrow 2H_{2\,(gas)} + O_{2\,(gas)} + 4e^-}$$

The drive of this bidirectional reaction depends of similar variables observed in the great majority of the chemicals reactions as temperature, pressure, and light intensity, reactants, by-products and products concentrations, as other known and unknown factors. These mean that under certain circumstances the address of cycle could be to the right or the left therefore it is not symmetrical in time.

Energetically, the most expensive part of the process is the breaking of the water molecule, which in the laboratory requires 2000 °C of temperature, however, the melanin contained normally in every eukaryotic cell, human cells included; takes a time of 3×10^{-12} seconds to harvest enough energy from the full electromagnetic spectrum to split, break, or dissociate the water molecule, which main product is molecular Hydrogen (H_2), the energy-carrier by excellence in Nature at the whole Universe, and inside our body can't be different, the other byproduct, diatomic Oxygen (O_2), is toxic at any concentration, at any level; so molecular Hydrogen is the element with the real value in this previously unknown reaction in humans. The splitting of the water molecule produces diatomic forms of both elements, Hydrogen and Oxygen.

Therefore, when the water molecule is dissociated, the energy released is carried by molecular Hydrogen, the smallest atom and at the same time, the energy carrier per excellence in the Universe. The second step in the Solis-Herrera Cycle is the water reformation that is, from the energetic point of view; the cheapest part, and the back-bonding of the previously released diatomic elements gives again water and an orderly flow of high energy electrons, four high energy electrons for each two reformed water molecules. The latter is other kind of energy different from the first one that is released and then carried by Hydrogen, so in both parts of the cycle should be available energy both of different characteristics, but finally available chemical energy.

We can image the way melanin release energy symmetrically in all directions, as alternating growing spheres of gas (H_2) and liquid (re-formed water) resulted from the breaking of the water molecule followed by a growing sphere of re-formed water that permeates constantly throughout the entire intracellular environment with these two different types of energy, taking relentless energy as all the cytoplasm and organelles contained therein, cell nucleus included. Hydrogen, with its precious load of energy, is used in many ways by the myriads of biochemical reactions that constantly happen within the cell. Recall that every single biochemical reaction has, as an unavoidable very first requirement: energy of activation; the small quantum of energy carried by Hydrogen are very useful for the cell metabolism because the hundreds or thousands of reactions that are simultaneously occurring into the cell are driven by the energy provided by this unsuspected source. In general terms, any biochemical reaction requires one or other type of "pushing" to occur, and it does not matter that the reaction could be classified as endergonic or exergonic.

Summarizing when the water of the CSF is dissociated by the neuromelanin of choroid plexuses or substantia nigra, or other pigmented places as locus coeruleus (the blue spot) or nucleus pigmentosus pontis; then molecular Hydrogen in gas form (H_2) is released, and thereafter the opposed reaction, the water reformation; then going to be a releasing of this reconstituted water beside a flow of electrons with enough characteristics as quantity, direction, so can be considered such energy. As any other chemical or biochemical reaction, the Solis-Herrera Cycle naturally trends to balance, however, the reactant and products concentrations, pressure, temperature, amount of visible and invisible light, *etc*. are determining factors for the process′ equilibrium. Translated to the ventricle size, when the balance is lost, it could be production of water and hence the size of the ventricles will be affected in one or another way. When there is a diminished rate of water dissociation, then they're going to be a greater amount of water, accordingly the ventricles will be enlarged and it is accompanied by unpredictable functional alterations at cell and brain level because the source of the 99% of the energy that every single cell needs is impaired, even more this alteration in the availability of energy is not uniform along the brain tissue, and could be changing frequently. Neuron cell requires energy for each process that we can assume and even to keep the shape of every single organelle, microtubule, cell membrane, mitochondria, Golgi apparatus, nucleus, *etc*. besides other important functions such as provide energy to synthesis of practically any compound, axoplasmic transport or flow, and the steady state and so forth. Human photosynthesis could be slowed down by one or another reason resulting in tissue changes that can be at molecular level only or if the decline in the turnover rate of water dissociation has enough duration then the distortion in the appearance of the tissue or its general characteristics will have modified enough to be interpreted as cortical atrophy [18], but we must keep in mind that first is the energy failure that induces metabolic abnormalities and second the anatomical changes (Fig. **2**).

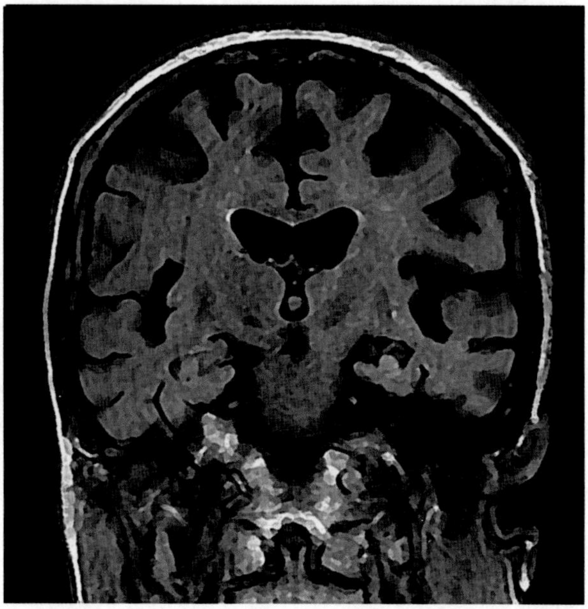

Fig. (2). Drawing tries to recreate the way in which the dissociation of the water molecule, molecular hydrogen, which is carried out by neuro-melanin; the energy is released by way of growing spheres (blue circles) that follow the laws of simple diffusion. When melanin is damaged by pollutants, such as pesticides, herbicides, fertilizers, metals, plastics, medicines, *etc*., the growing sphere of energy is expected to be smaller in size and scope; Therefore, it covers less area of brain tissue, which, without main source of energy, tends to deteriorate and lose capabilities.

Because energy distribution becomes uneven the cell´s behavior and hence clinical manifestations in accordance with magnitude of tissue alterations derivative of the damage in the very first spark of life or the real engine of neuron-cell functions that is the water dissociation or Solis-Herrera Cycle. Therefore, the clinical picture will be congruous with the suffered damage.

Studies have demonstrated a large proportion of persons with schizophrenia have profound cognitive deficits. These include difficulties with higher-order attention, insight, judgment, abstraction abilities, comprehension and retention of complex verbal and non-verbal material, and recognition and production of non-verbal affects (*e.g.* the reading of emotional expression on a face, the production of emotional sounds/nuances when speaking). The pattern of deficits is often unique to an individual, although patterns across sub-groups of persons with schizophrenia are seen.

Our findings of the unsuspected intrinsic property of melanin to split the water molecule can explain very well the ancient observation of the ventricle enlargement in several mental illnesses [19]. The former explanation that the increase in the size of ventricles in regards to developmental delays and/or brain tissue atrophy with the subsequent passive filling of the empty space by the CSF should be superseded.

The right balance of the reaction and therefore the correct ventricle size is grounded in this fascinating process:

$$\mathbf{2H_2O\ (liquid) \leftrightarrow 2H_{2\,(gas)} + O_{2\,(gas)} + 4e^-}$$

When the Solis-Herrera Cycle has an adequate equilibrium or $\Delta G = 0$, means that both reactions, dissociation/re-formation, occur at the same rate; thereby the amount of water or CSF is correct, and hence the ventricular size will be the appropriate, by other side, the energy released in the first step of the cycle or the water dissociation, which requires higher amounts of energy than the second step, going to be in the right levels in regards tissue energy needs for multiple metabolic aims [20]. In the righteous circumstances of light, water, amount and nature of the metabolic intermediate, pressure, humidity, electrolytes; *etc*. then we will have enough quantity of energy for the neuron cell, appropriate amount of water or CSF, normal ventricular size, and right cortical thickness. Mental functions under these circumstances will be proper.

However, imbalance could be skewed in several ways:

$$\mathbf{2H_2 \leftrightarrow O\ 2H_2 + O_2 + 4e^-}$$

In this example, we could infer that water is being consuming quickly, and hence dehydration of tissue ensues.

$$\mathbf{2H_2 \leftrightarrow O\ 2H_2 + O_2 + 4e^-}$$

The deviation of the equation right it's most difficult to occur, because widely available diatomic Hydrogen and Oxygen are required, and the only source of these molecules in such quantities is precisely the dissociation of the water molecule.

In the case of exogenous administration of Oxygen, *i.e*. medical aims, then we will have:

$$\mathbf{2H_2 \leftrightarrow O\ 2H_2 + O_2 + 4e^-}$$

This therapeutic measure instead to improve the balance worsening the Hydrogen availability, because accelerated water reformation is induced by the abnormally highest levels of exogenous Oxygen, that is toxic, finally the tissue energy levels will be comprised per hydrogen consumption. Tissues will have edema or ventricle enlargement.

Fig. (**2**) shows the increasing spheres of chemical energy that is transported mainly by molecular hydrogen because high energy electrons are considered to have little penetration because they are absorbed almost immediately, besides that, when traveling at the speed of light, they are not controlled easily, so the energy that transports the molecular hydrogen is very assimilated by the cellular structures. In general terms, the fibrous part of the cell, that is: the fixed organelles; they take advantage of the hydrogen and the high energy electrons, and the soluble parts of the cell, those that do not have a fixed placement, mainly use the molecular hydrogen.

CONCLUSION

Ancient dogma that the glucose is a molecule that is so wonderful that it can provide the energy required for its own metabolism, is now broken into a thousand pieces. From the thermodynamic point of view is not possible.

There are both anatomic and functional brain alterations that can be observed when glucose in plasma levels decrease or increase, and are due to our body, as well as the entire universe, is in a delicate balance between energy and mass.

The glucose is only source of carbon chains, with which, the body cells synthesize 99% of biomolecules that conform the biomass body. Eukaryote cell knows and drive glucose from the beginning of time, therefore handled it very well, because it is stretched, twisting it, fragmented it precisely, and combines it with other elements such as nitrogen, different carbon chains and carbon atoms; minerals; *etc*. But at the end it is only biomass.

The energy that is required for this purpose, defined as everything that produces a change, our body takes from water, like plants, although we know that it comes strictly from the light, but it must be through the dissociation of the water molecule.

If the glucose were energy supply, then diabetic patients would fly. And it is not so, in fact, diabetic patients suffer neurological damage, functional and anatomic; that is comparatively more severe than patients who are not diabetics, for example, in the case of cerebral vascular events.

On the other hand, grooves or brain caesuras, now will be considered differently, as it is the way in which nature ensures that the neurons are in direct contact all the time with its source of energy which is the water of the ventricles and subarachnoid space.

The locus coeruleus (LC) or nucleus pigmentosus pontis (Fig. **3**), meaning "heavily pigmented nucleus of the pons", contains neuromelanin that is formed by the polymerization of noradrenaline and is analogous to the black dopamine-based neuromelanin in the substantia nigra. The locus coeruleus function may figure in clinical depression, panic disorder, anxiety and posttraumatic stress disorder (PTSD).

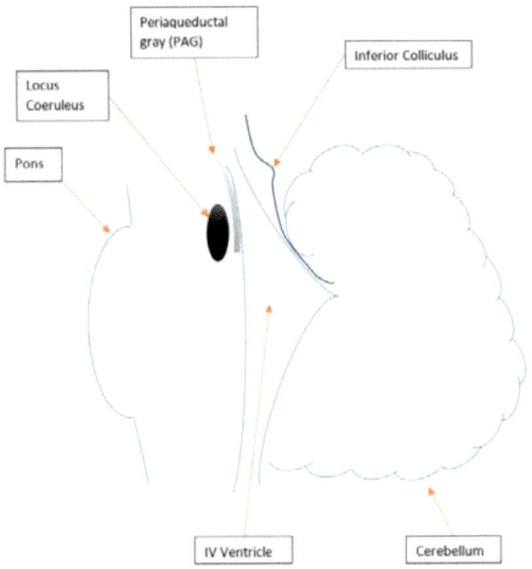

Fig. (3). Approximate location of locus coeruleus. It is expected that zones with high metabolism are placed around pigmented areas, especially in upper parts, due to molecular hydrogen (H_2) is lightest and trend to goes upward.

It is conceivable, that the melanin, which is its origin or location, have the same function: transforming the visible and invisible light into chemical energy through the dissociation of the molecule of water, as the chlorophyll in plants.

There is up to 70% loss of locus coeruleus neurons in Alzheimer´s disease [21], means available chemical energy (from neuromelanin) that are impoverished; thereby neuron cells cannot make regular functions.

HUMAN AND ANIMAL RIGHTS

No humans were used in this research. All animal research procedures followed were in accordance with the standards set forth in the eighth edition of Guide for the Care and Use of Laboratory Animals published by the National Academy of Sciences, The National Academies Press, Washington, D.C.

CONFLICT OF INTEREST

The author (editor) declares no conflict of interest, financial or otherwise.

ACKNOWLEDGEMENT

We are grateful to the Human Photosynthesis® Study Center, its continuous and unconditional support.

REFERENCES

[1] van Boxel, P.; Bridges, P.K.; Bartlett, J.R.; Trauer, T. Size of cerebral ventricles in 66 psychiatric patients. *Br. J. Psychiatry,* **1978**, *133*, 500-506. [http://dx.doi.org/10.1192/bjp.133.6.500] [PMID: 310699]

[2] Dandy, W.E. Roentgenography of the brain after injection of air into the spinal canal. *Ann. Surg.,* **1919**, *70*(4), 397-403. [http://dx.doi.org/10.1097/00000658-191910000-00004] [PMID: 17864170]

[3] Moore, M.T.; Nathan, D.; Elliot, A.R.; Laubach, C. Encephalographic studies in mental disease. *Am. J. Psychol.,* **1935**, *92*, 43-67. [http://dx.doi.org/10.1176/ajp.92.1.43]

[4] Haug, J.O. Pneumoencephalographic studies in mental disease. *Acta Psychiatr. Scand. Suppl.,* **1962**, *38*(165), 1-104. [PMID: 13960992]

[5] Lewis, S.W. Computerised tomography in schizophrenia 15 years on. *Br. J. Psychiatry Suppl.,* **1990**, (9), 16-24. [PMID: 2132566]

[6] Internet Brain Volume Database. http://www.cma.mgh.harvard.edu/ibvd

[7] Harrison, P.J. The neuropathology of schizophrenia. A critical review of the data and their interpretation. *Brain,* **1999**, *122*(Pt 4), 593-624. [http://dx.doi.org/10.1093/brain/122.4.593] [PMID: 10219775]

[8] Van Horn, J.D.; McManus, I.C. Ventricular enlargement in schizophrenia. A meta-analysis of studies of the ventricle:brain ratio (VBR). *Br. J. Psychiatry,* **1992**, *160*, 687-697. [http://dx.doi.org/10.1192/bjp.160.5.687] [PMID: 1534268]

[9] Raz, S.; Raz, N.; Bigler, E.D. Ventriculomegaly in schizophrenia: is the choice of controls important? *Psychiatry Res.,* **1988**, *24*(1), 71-77. [http://dx.doi.org/10.1016/0165-1781(88)90142-4] [PMID: 3293096]

[10] Carpenter, W.T., Jr; Buchanan, R.W. Schizophrenia. *N. Engl. J. Med.,* **1994**, *330*(10), 681-690. [http://dx.doi.org/10.1056/NEJM199403103301006] [PMID: 8107719]

[11] Lin, K-M.; Kleinman, A.M. Psychopathology and clinical course of schizophrenia: a cross-cultural perspective. *Schizophr. Bull.,* **1988**, *14*(4), 555-567. [http://dx.doi.org/10.1093/schbul/14.4.555] [PMID: 3064282]

[12] Davis, K.L.; Kahn, R.S.; Ko, G.; Davidson, M. Dopamine in schizophrenia: a review and reconceptualization. *Am. J. Psychiatry,* **1991**, *148*(11), 1474-1486. [http://dx.doi.org/10.1176/ajp.148.11.1474] [PMID: 1681750]

[13] Shelton, R.C.; Weinberger, D.R. Brain Morphology in schizophrenia.*Psychopharmacology: the third generation of progress*; Meltzer, H.Y., Ed.; Raven Press: New York, **1987**, pp. 773-781.

[14] Shenton, M.E.; Kikinis, R.; Jolesz, F.A.; Pollak, S.D.; LeMay, M.; Wible, C.G.; Hokama, H.; Martin, J.; Metcalf, D.; Coleman, M. Abnormalities of the left temporal lobe and thought disorder in schizophrenia. A quantitative magnetic resonance imaging study. *N. Engl. J. Med.,* **1992**, *327*(9), 604-612.
[http://dx.doi.org/10.1056/NEJM199208273270905] [PMID: 1640954]

[15] Buchsbaum, M.S.; Haier, R.J.; Potkin, S.G.; Nuechterlein, K.; Bracha, H.S.; Katz, M.; Lohr, J.; Wu, J.; Lottenberg, S.; Jerabek, P.A. Frontostriatal disorder of cerebral metabolism in never-medicated schizophrenics. *Arch. Gen. Psychiatry,* **1992**, *49*(12), 935-942.
[http://dx.doi.org/10.1001/archpsyc.1992.01820120023005] [PMID: 1360198]

[16] Dominic, F.; Lakshika, T.; Alex, O.; Seamus, O. Third ventricle enlargement and developmental delay in first-episode psychosis: preliminary findings. *Brit. J. Psych.* **2000** 177: 354-359 doi: 10.1192/bjp.177.4.354.

[17] Solís-Herrera, A.; Arias-Esparza, M de C.; Solís-Arias, R.I.; Solís-Arias, P.E.; Solís-Arias, M.P. The unexpected capacity of melanin to dissociate the water molecule fills the gap between the life before and after ATP. *Biomed. Res.,* **2010**, *21*(2), 222-225.

[18] Solís-Herrera, A.; Arias-Esparza, Ma del C. The Unexpected capability of melanin to split the water molecule and the alzheimer´s disease. *Neurosci. Med.,* **2011**, *2*, 217-221.

[19] Soares, J.C.; Mann, J.J. The anatomy of mood disorders--review of structural neuroimaging studies. *Biol. Psychiatry,* **1997**, *41*(1), 86-106.
[http://dx.doi.org/10.1016/S0006-3223(96)00006-6] [PMID: 8988799]

[20] Solís-Herrera, A. The pharmacologic intensification of the water dissociation (Human Photosynthesis) and its effect over tissues affected by bloodshed of diverse etiology. *Int. J. Clin. Med.,* **2011**, *2*(2), 332-338.
[http://dx.doi.org/10.4236/ijcm.2011.23058]

[21] Bondareff, W.; Mountjoy, C.Q.; Roth, M. Loss of neurons of origin of the adrenergic projection to cerebral cortex (nucleus locus ceruleus) in senile dementia. *Neurology,* **1982**, *32*(2), 164-168.
[http://dx.doi.org/10.1212/WNL.32.2.164] [PMID: 7198741]

CHAPTER 7

Age-related Macular Degeneration, the Alzheimer´s Disease of the Eye and the Intrinsic Property of Melanin to Transform Light Power into Chemical Energy

Arturo Solís Herrera*, **Ruth I. Solís-Arias, Paola E. Solís-Arias** and **Martha P. Solís-Arias**

* *Human Photosynthesis® Study Center, Av. Aguascalientes Norte 607, Pulgas Pandas Sur, CP 20230, Aguascalientes, México*

Abstract:
Background: Age is a common risk factor for Alzheimer's disease (AD) and age-related macular degeneration (AMD). In addition to age as a risk factor, AD and AMD share several pathological hallmarks including β-amyloid (Aβ) accumulation, oxidative stress, and apoptotic cell death. An important characteristic common to both diseases is the presence of amyloid β (Aβ) in the senile plaques of the AD brain and in the drusen of AMD patients. Age-related macular degeneration (AMD) is the first cause of blindness in cold countries since five decades ago or more among people 55 years of age or older. According to one estimate in 2006, 29% of the United States population, 75 years of age and older, has signs of AMD [1]. Statistical behavior of this macular disease has not changed in a significant way in spite of different types of available treatments as photodynamic therapy, biological agents anti VEGF, and vitamin supplements. However, amyloid beta had been described in AMD since 1982.

Methods: Patients with AMD and AD in several stages of evolution were enrolled in this study. The study was approved by Ethic committee at our Human Photosynthesis® Study Center. A novel therapeutic approach is used in this study based on the discovery of the Human Photosynthesis® and the treatment is primarily the medical enhancement of this very primary process in the sequence of physiological events. This work shows the therapeutic results and examples of patients with AMD and AD treated under this thesis, suggesting that a low level of Human Photosynthesis® is a constant subjacent cause, but indeed a very important one. It is expected that if AMD improves dramatically in spite of the presence of edema, hemorrhage, neovascularization, fibrosis and amyloid beta; Alzheimer´s disease can also be improved since AD and AMD have similar characteristics.

* **Corresponding author Arturo Solís Herrera:** Human Photosynthesis® Study Center, Av. Aguascalientes Norte 607, Pulgas Pandas Sur, Aguascalientes, México; C.P. 20138; Tel/Fax: +524492517232; E-mail: comagua2000@yahoo.com

Results: Anatomical and functional results in AD and AMD are very encouraging. Improvement in visual function is highly significant, around 50-80% in average. On the other hand, also the AD patients treated with enhancement of Human Photosynthesis® therapy showed important regression in their signs and symptoms.

Conclusion: A common pathogenic mechanism might exist between AMD and AD. Thus, therapeutic approaches which have targeted Aβ in patients with AD can also be applied to AMD. The enhancement of the Human Photosynthesis® in patients affected with AMD and AD is a novel promissory treatment and highly effective.

Keywords: AMD, Amyloid β, AD, Energy, Human Photosynthesis®, Macula, Melanin, Monoclonal antibodies, Retina.

BACKGROUND

Alzheimer Disease (AD), described for the first time by Alois Alzheimer in 1901, is a progressive, irreversible, neurodegenerative disease of unknown etiology that has been reported to affect more than 5.3 million people in the United States. By 2050, the number of individuals with AD could range from 11.3 million to 16 million (Hebert, Scherr, Bienias, *et al.*, 2003).

At present, few agents that are approved by the US Food and Drug Administration for the treatment of AD have demonstrated only modest effects (if any) in modifying clinical symptoms for relative short periods; none has shown a clear effect on disease progression [2]. Statistically, the effect of approved drugs is significant, but clinically is null, thereby, new therapeutic approaches are desperately needed (Grammas *et al.*, 2011).

Age-related Macular Degeneration (AMD) is a late-onset, neurodegenerative retinal disease that shares several clinical and pathological features with Alzheimer´s disease (AD) including extracellular deposits containing Amyloid beta [3]. AMD is manifested as progressive loss of central vision due to dysfunction and death of photoreceptor and adjacent retinal pigment epithelial cells (RPE) at the macula. It is the leading cause of irreversible vision loss in people over the age of 65 [4]. Women were found to have a slightly higher incidence of AMD than men, and white Americans were found to have fivefold-to-six fold higher ascertainment criteria than black Americans. It is a very well-known fact that there are similarities between AMD and another neurodegenerative disease [5].

The neo-vascular (wet) type of AMD is less prevalent than atrophic (dry) type of AMD, and the former accounts for most severe visual loss from AMD. The wet type of AMD is characterized by the presence of choroidal angiogenesis accompanied by fibrous tissue. Angiogenesis is the process by which new vessels are created from pre-existing normal vasculature. Increased rates of angiogenesis are associated with several disease states, including cancer, psoriasis, rheumatoid arthritis, and diabetic retinopathy.

The vascular endothelial growth factor (VEGF) seems to be a key contributory factor in the patho-physiology underlying neo-vascular AMD; but, the recent development in anti-VEGF therapies, although apparently a logical approach; has unfortunately demonstrated short life and limited efficacy, besides this, many intra-vitreal injections are often required, with the serious risks inherent to the visual function that a repetitive intraocular injection carries and worse still and they do not improve at all the cases of wet AMD [6].

AMD is a multifactorial disease in which the relative contributions of genetic and environmental factors remain unknown. The strongest known risk factor for AMD is advanced age. Like AMD, AD is also a multifactorial disease affected by age, environmental factors and genetics [7]. The hallmark of AD considered so far includes the presence of senile plaques, neurofibrillary tangles, neuronal cell loss, reactive gliosis, and in some cases cerebrovascular amyloid deposits. Senile plaques are generated by deposition and accumulation of the amyloid beta peptides. Therefore, two of the most common neurodegenerative diseases, AD and AMD, could simultaneously be treated or prevented with a single therapeutic intervention (Ding, Ling, Mace, Herrmann, Sullivan and Bowes Rickman 2007).

Brain micro-vessels cultured from patients with Alzheimer disease (**AD**) were found to release a myriad of factors that have been implicated in vascular activation and angiogenesis [8]. The signaling cascades associated with vascular activation and angiogenesis are up-regulated in AD-derived brain micro-vessels. Available data in the literature are consistent with the observation that factors and processes characteristic of vascular activation and angiogenesis are found in the AD brain, however, there is a lack of evidence for increased vascularity in AD. Cerebral hypo-perfusion resulting in, hypoxia, which constitutes one of the most potent stimuli for irreversible activation of vascular endothelial cells. In working models, hypoxia stimulates the angiogenic process in diverse tissues as cornea, retina and neuronal tissue.

The order of the processes that finally impair photoreceptors function in AMD or neurons in AD is not well understood. There are many risk factors as atherosclerosis, cardiovascular disease association; smoking, alcoholism; iris color, skin sun sensitivity; ethnicity; solar radiation, sun exposure; *etc.* Cellular aging processes have similar organelle and signaling association in the retina and brain tissues [9]. With regard to anatomical factors in retina and choroid, macular drusen in their different types and forms are considered as characteristic clinical feature of AMD. Amyloid β, a major extracellular deposit in Alzheimer´s disease plaques has recently been found in drusen from AMD, the hallmark extracellular deposit in AMD [10]. The presence of large drusen or confluent drusen as well as drusen types, such as soft and indistinct, has been recognized as risk factors for the progression of advanced AMD, which includes central geographic atrophy (GA) and choroidal neovascularization [11].

The angiogenic process is complex and involves several known and unknown steps. Stimuli to initiate angiogenesis, include hypoxia, inflammation and mechanical factors such as shear stress and stretch. VEGF promotes endothelial survival, proliferation, and migration through numerous pathways. VEGF takes part in pathological growing (Angiogenesis) and the normal growing, remodel and maintenance of the healthy blood vessels, for instance in fetus (Vasculogenesis). Therefore, anti-VEGF therapy could be harmful under certain circumstances due to the inactivation of this important factor which can cause damage in otherwise healthy tissues by interference with the physiology of normal blood vessels and thereby, sooner or later, the affected tissues will become a part of the disease.

The issue of pro-angiogenic and anti-angiogenic signals in AD and AMD is complex. In AD brain, despite increases in several pro-angiogenic factors, there is still lack of evidence for increased vascularity. However, cerebral hypo-perfusion and decreased micro-vascular density are major clinical features in AD. In spite of the continued presence of pro-angiogenic stimulus, it seems that perhaps an ever changing imbalance of pro-angiogenic and anti-angiogenic factors or aborted angiogenic signaling (by poorly understood reasons) prevents new vessel growth or at least the maintenance of a healthy vascular tree in the AD brain. In both AD and AMD diseases, the detrimental intra- and extracellular deposits have many similarities. Cellular aging processes have similar organelle and signaling association in the retina and brain tissues. For instance, in the dry type of AMD, there is retinal and choroidal hypo-perfusion and decreased micro-vascular density, as described in AD brain.

The Photosynthesis in Humans or Melanin´s Molecule Intrinsic Property to Splits and Reform the Water Molecule

The steady state of any cell, tissue; or organ in our body is a set of highly complex biochemical processes that are beyond our understanding because they are evolved based on 4 billion years of evolution. So far the replication of a single human hair in the laboratory is not yet possible. However, our finding of the intrinsic property of melanin to split and reform the water molecule breaks the ground [12].

In 1990, a study about the three main cause of blindness was conducted with limited facilities because their incidence and prevalence have not changed since 50 years ago or more. Therefore, available treatments were not working at all. That was the main reason to start a descriptive study on the live patients about Glaucoma, Diabetic retinopathy and Age-Related Macular Disease.

The tested hypothesis was based, at that time, on a prevalent theory that blood vessels are able to fulfill any metabolic tissue requirement. Thereby a careful observation of the blood vessels in and around the optic nerve could give early clues of the cited illnesses, therefore, allowing early treatments. The first challenge was the small diameter of the optic nerve in the human live patient, which was about 1200 microns, or twelve human hair altogether. So to reach an adequate magnification of this key structure and the modification of several optical characteristics in the design of standard examination equipment were necessary.

The main variable in the study was characteristics of the minute blood vessels of the optic nerve. However, blood vessels are not able to transport energy, at most they can only carry building blocks as glucose, amino acids; lipids, electrolytes; *etc.*, substances that eventually intervene in endergonic reactions. Thereby, food is only a source of biomass′, and thereby cannot be a source of energy. With meals, the body makes skin, bone, muscle, neuronal tissue, cell membrane, organelles; hormones, *etc.* But the energy or the force that the cells need and therefore the body needs to imbue life to the relatively inanimate ingested substances at meals is taken from water.

Studies about retinal blood vessels have been done exhaustively since decades, but with disappointing results. And the proof is that the Age-related macular disease cannot be changed as the first cause of blindness in cold countries and the third one in warm countries until today.

In 1990, during the observational study about the three main causes of blindness in the world, it was observed that glucose is not the source of energy of retinal cells instead retinal tissue takes energy from water.

The chemical reaction by which retinal cells take energy from water is:

$$2H_2O \leftrightarrow 2H_2 + O_2 + 4e^-$$

It was an astonishing finding in humans because this type of reaction is only observed in plants, and in the leaves of the plants the reaction can be schematized as follows:

$$2H_2O \rightarrow 2H_2 + O_2$$

Due to their similarities, the reaction in humans was named as Solís-Herrera Cycle® (by its discoverer) or Human Photosynthesis® in order to draw the attention that the very first reaction in plants and humans are practically the same.

It can be said that life in plants and humans begin with the same step: the dissociation of the water molecule in diatomic hydrogen and diatomic oxygen.

The energy released by this process is transported by the diatomic hydrogen, the energy carrier by excellence in the entire universe, therefore plants and humans cannot be different.

This chemical energy is used by the cell to impel the consequent reactions that *in total* mold the self-sustainable chemical system, which eventually goes into Darwinian evolution or in other words the life itself.

The therapeutic approach used in this study is based on our finding of the analogue first reaction in humans and plants that can be schematized as::

$$\mathbf{2H_2O \leftrightarrow 2H_2 + O_2 + 4e^-}$$

In humans (see above) the water molecule is dissociated and reformed.

$$\mathbf{2H_2O \rightarrow 2H_2 + O_2}$$

However, in the leaves of the plants, the reaction takes place in one way, water molecule can be broken down but cannot be reformed and plants expelled the O_2 to the atmosphere.

It is very interesting that the main product of the water dissociation is diatomic hydrogen, being an excellent energy carrier in the universe, not only of human body. Whereas, diatomic oxygen is a stable molecule; but it is toxic at any concentration.

On the basis of the above discussed arguments it can be said that the very first reaction in plants and in humans is almost the same, and by analogy the reaction observed in humans was named by the research team as Human Photosynthesis®. This thesis is based on evaluating that if we enhance the water dissociation and its consequent back-bonding that are parts of a cycle named as Solis-Herrera´s cycle®, then the balance of the consequent biochemical processes that was gradually hanging of this very first spark of life along four billion years of evolution will be restored, and in AD and also in AMD patients [13], whether the improvement will be huge due to the unknown number, sequence; temporality; spatial arrangement; and order of the biochemical reactions involved in the maintenance of healthy tissues and their components would be recovered. Once the balance between energy and biomass is regained in the involved tissues in AD and AMD patients, the body is then able to keep the adequate shape and function as it has done millions of years, millions of times.

A normal tissue requires same characteristics of the involved multiple and highly complex bioprocesses such as; sequence, temporality, spatial arrangement and order as occuring naturally and has been gradually developed and implemented along four billion years of evolution. More importantly, any chemical or biochemical reaction has a mandatory first requirement: available energy and the entire human body have no exceptions.

Human Photosynthesis® teaches us that glucose cannot be a source of energy. It is only an important metabolic source. Therefore, food can only be a source of biomass′, and thereby cannot be a source of energy. With meals, human body makes skin, bone, muscle, neuronal tissue, cell membrane, organelles; hormones, *etc.* But the energy or the force that the cells and our entire body needs to imbue life to the relatively inanimate ingested substances is taken from water. Until today, it is firmly believed (wrongly) that any metabolic tissue's requirement is fulfilled through the blood stream. However, blood vessels are not able to transport energy, at most they only carry building blocks as glucose, amino acids; lipids, electrolytes; *etc.*, substances that eventually intervene in endergonic reactions.

The real source of energy of human eukaryotic cell is the sunlight, the universal nutrient. Human eukaryotic cell has unsuspected capacity to convert photonic into chemical energy by means of the hitherto unknown Human Photosynthesis® system composed of Light (visible and invisible)/Melanin/Water, arranged in order of abundance in the universe. Probably it is not only limited to humans; perhaps it is the universal photosynthesis system.

In an attempt to explain the therapeutic proposal, it can be said in other words that the origin and life′s support of the body is photosynthesis. Any biochemical reaction in any cell, or; any intracellular organelle has its origin in this very first reaction. In the words of Darwin: "the very first spark of life". Thereby, the human body comes from and depends completely on one or another way of this very first reaction, the Solis-Herrera Cycle®; or the intrinsic capacity of melanin to split and reform the water molecule and the reaction is represented as follows:

$$2H_2O \leftrightarrow 2H_2 + O_2 + 4e^-$$

Therefore, if this astonishing process is intensified medically, the body undoubtedly will respond as a whole. The improvement will be noted in joints, heart, lung; kidney, intestine, brain, eye; *etc.* The AMD eye and AD brain will be better because the entire body also improves. Thereby, in the aging population, two of the most common neurodegenerative diseases, AD and AMD, could simultaneously be treated or prevented with a single therapeutic agent [14].

Material and Methods

A total 7 groups of 10 male wistar rats were formed, one of them was the control group and the other six groups were exposed to different concentrations of a previously identified compound (C_6H_5OH) with the intrinsic property to induce a marked down turn in the photosynthesis activity in mammals and, therefore, in humans. The C_6H_5OH compound in aqueous solution was applied topically in the right eye cornea of all the animals at different concentrations and the left eye was used as control. The theoretical framework suggested that when the photo-dissociative activity of melanin down regulates, blood vessels tend to grow abnormally, thus, as expected; eight of every 10 corneas, in average; of each exposed group developed diverse degrees of neovascularization which were concentration-dependent. The toxic compound was applied just one time in all cases. Sixty-five percent of the exposed corneas have angiogenesis fifteen days later. No left eye in any group developed neovascularization (Figs. **1-5**).

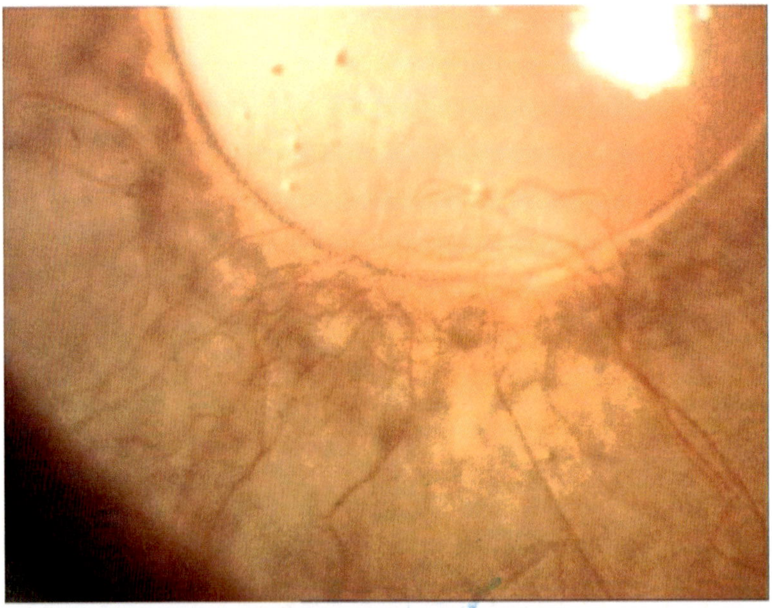

Fig. (1). Right Cornea of male Wistar rat, fifteen days after the exposure to just one drop of the toxic compound. The growth of the blood vessels on the surface of the cornea is easily noticeable with a very small fibrous component.

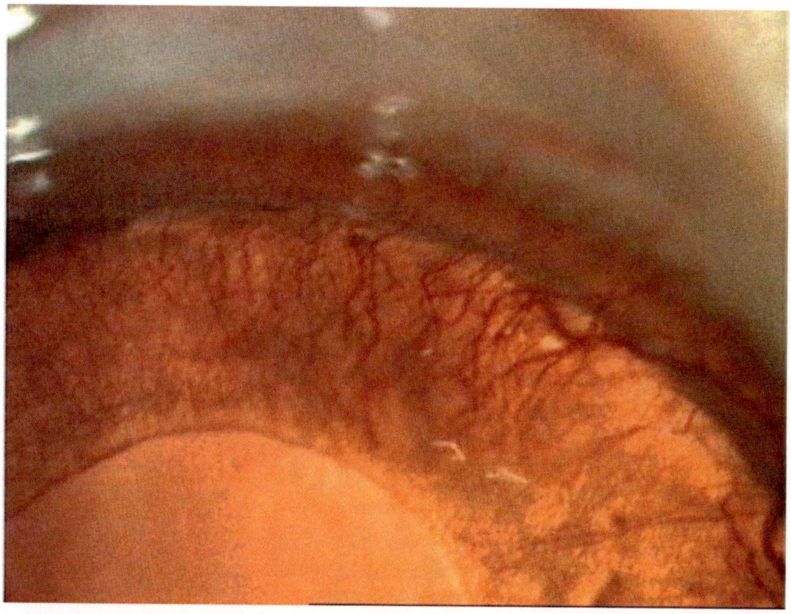

Fig. (2). Angiogenesis in cornea of male Wistar rat, group 5.

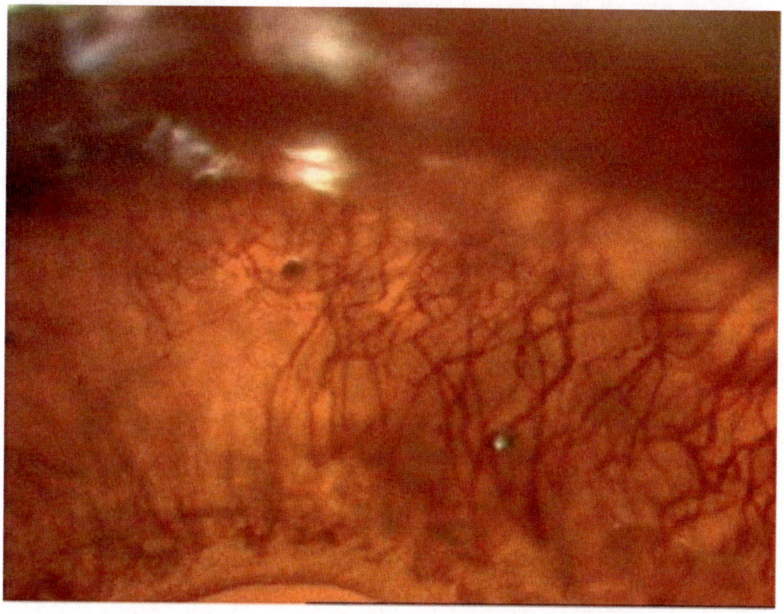

Fig. (3). Enlargement of the corneal blood vessels of the right eye of the male Wistar rat of Fig. (**2**).

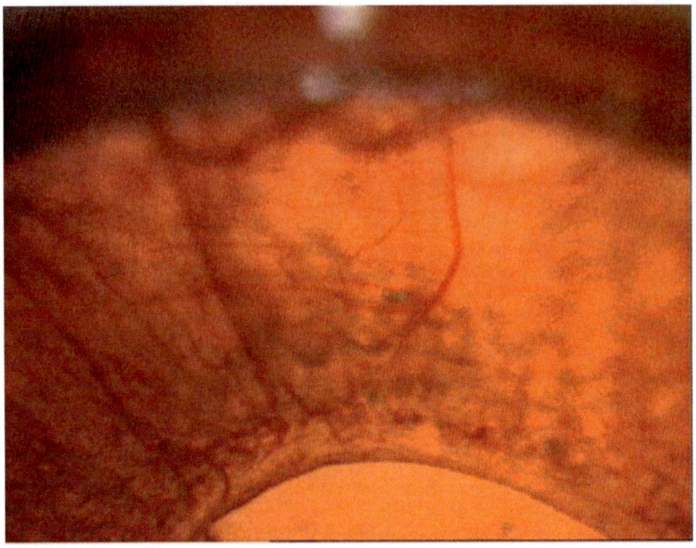

Fig. (4). Male Wistar rat, group 3. The photograph shows the cornea of the right eye; fifteen days after exposure. Angiogenesis is evident at the surface of the cornea.

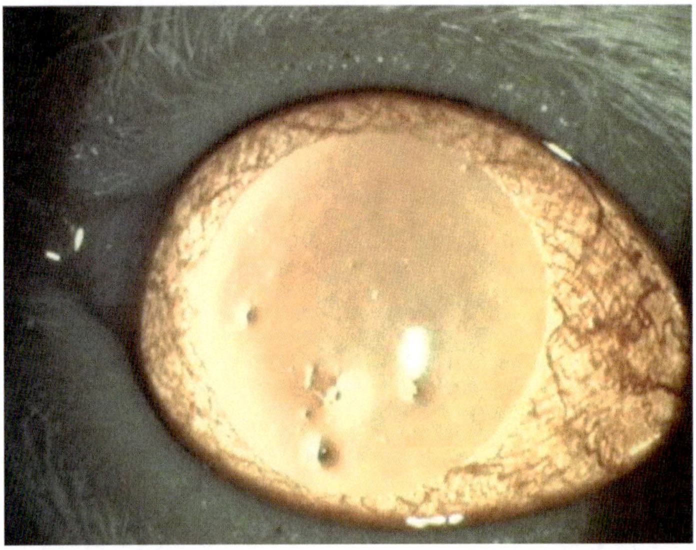

Fig. (5). The left eye of male wistar rat, group 2. There is no blood vessel at the surface of the cornea. The visible blood vessels in this photograph correspond to the normal appearance of the iris blood vessels.

The experimental results conformed to the predictions made. When Human Photosynthesis® is slowed down in acute form (cold, heavy contamination in air or water, anesthetic agents) then edema and hemorrhage ensue (Fig. **6**), however, if Human Photosynthesis® process is lowered in a chronic manner, then fibrosis or mitosis will be present in any tissue.

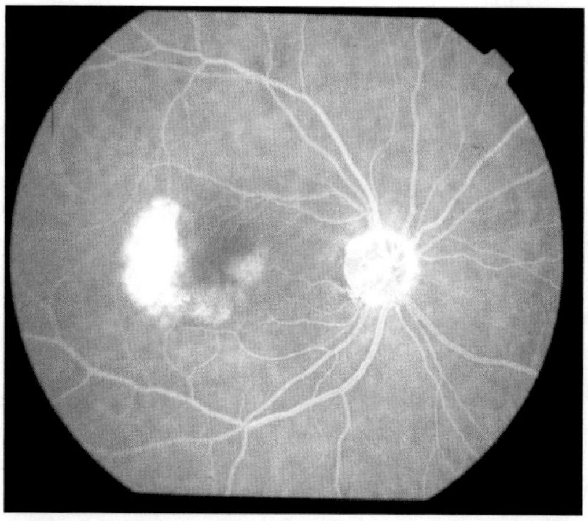

Fig. (6). Female patient, 82 years old, with progressive vision loss for four months of evolution. The patient has only a mild systemic arterial hypertension, and has no diabetic antecedents. In this photograph taken during the first day examination, the presence of sub-retinal membrane is manifested by the leaking of fluorescein. Vision at that time was 20/200. Treatment with the Human Photosynthesis® enhancer QIAPI 1™ (U.S. Pat. No. 9, 918, 996) was initiated at doses of three drops sublingually every two hours during day time.

Therapeutic Results in Age-related Macular Degeneration (Fig. 7 - 25)

Age-related macular disease has edema and hemorrhage at first and gradually abnormal fibrosis and mitosis will develop in the choroidal tissue with consequent progressive distortion in the anatomy of the overlying retina and adjoined choroid and pigment retinal epithelium; therefore, central vision is impaired, sooner or later; substantially.

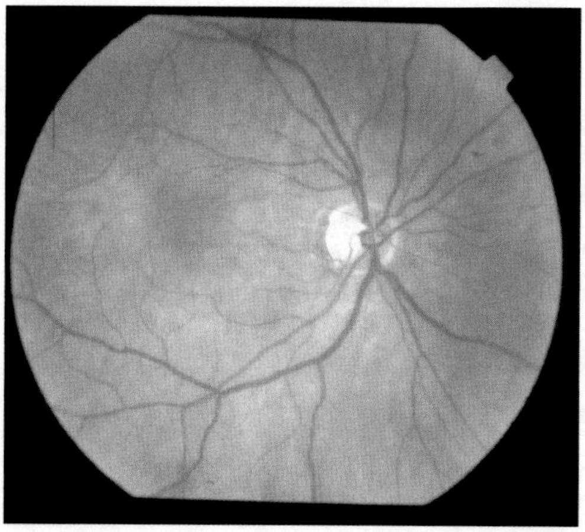

Fig. (7). The same case of the Fig. (6), the photograph was taken six weeks later; anatomic and functional improvement was dramatic. Recovery of the vision reached 20/25. The treatment with QIAPI 1™ (U.S. Pat. 9, 918, 996) was continued for several months.

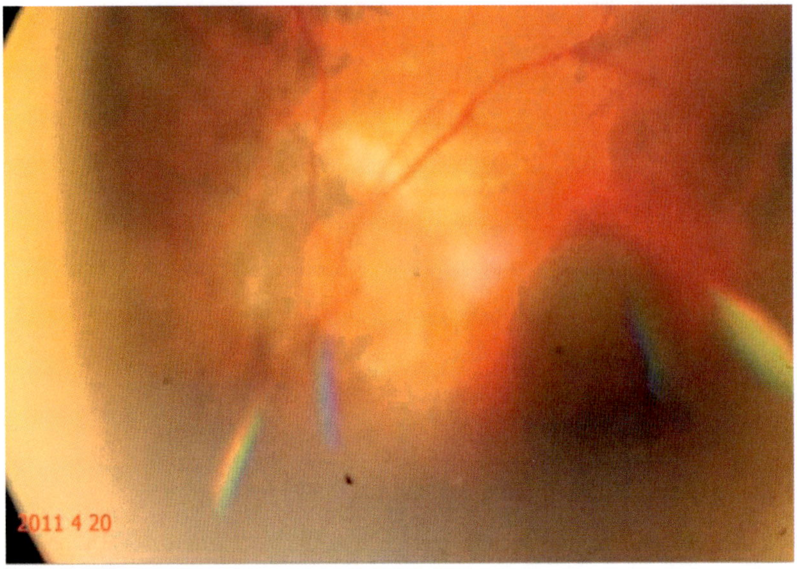

Fig. (8). Female patient, with 68 years of age; her myopia was in the range of -12.00 diopters, without systemic hypertension nor diabetes; the eye had an intraocular lens placed two and half years ago and the vision loss had eight months of evolution. This photograph was taken on the first day, during the examination that shows a sub-retinal membrane located at the macular area in the left eye with signs of recent bleeding. The enhancement of the Human Photosynthesis® with QIAPI 1™ (U.S. Pat. No. 9, 918, 996) was started at the same day at doses of three drops under lingua each two hours during day time.

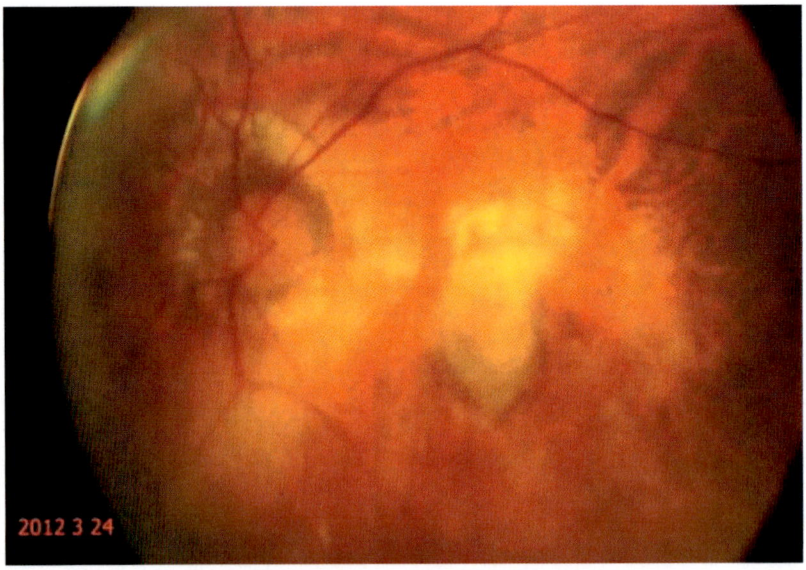

Fig. (9). Nine months later, the photograph shows a sub-retinal membrane with significant retraction and also improvement in the anatomic characteristics of the retinal and choroidal blood vessels.

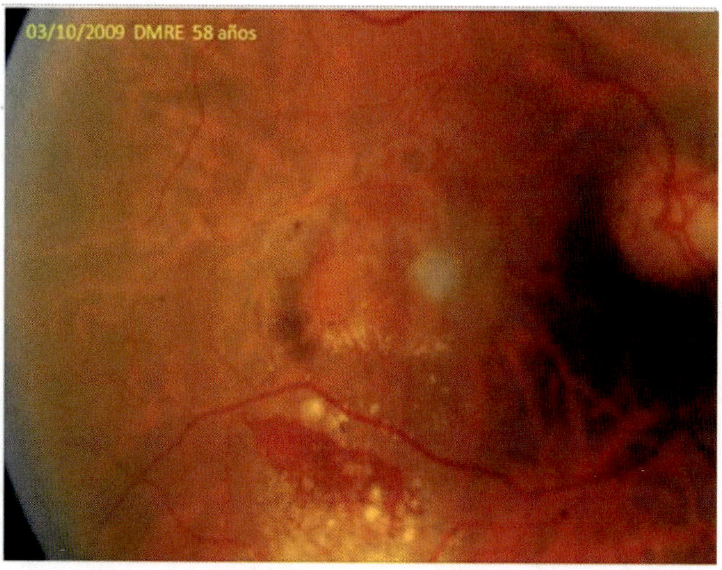

Fig. (10). Male patient, 58 years old, diagnosed with Parkinson disease since 2005 and AMD was diagnosed two years ago (2007). This was the clinical aspect of the sub retinal membrane during the first day of examination. Treatment was started with QIAPI 1™ (U.S. Pat. No. 9, 918, 996) at doses of three drops, sublingually each two hours during day time.

Fig. (11). Fourteen months later, retinal tissue shows less distortion, with no hemorrhages at all and the exudates now are placed in most peripheral sites. In spite of the fact that the patient has not yet completely recovered, the healthy appearance of retinal layer draws attention towards the fact that typical macular scarring has not developed during these two years of treatment, and the progressive improvement is the usual behavior of the tissue. The vision is 20/100.

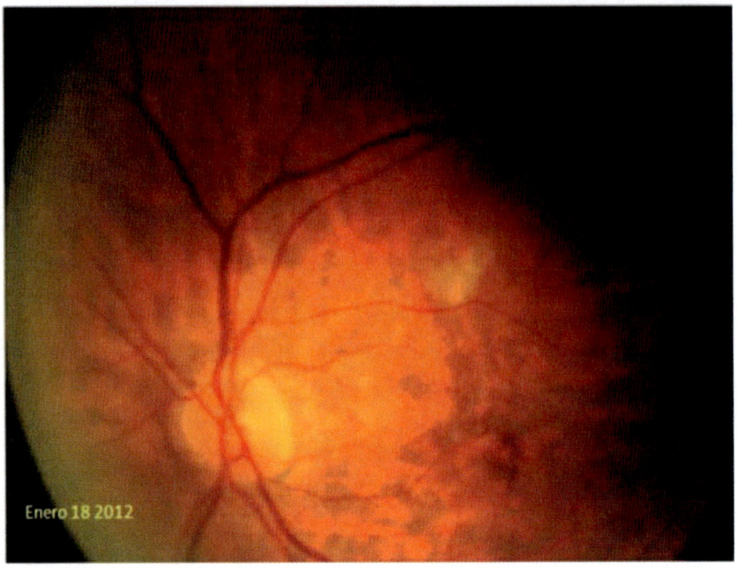

Fig. (12). A female patient, 61 years old; with vision loss of two months of evolution, with no hypertension and diabetes. The visual acuity diminished to 20/200 abruptly 15 days ago. There is a round shaped hemorrhage in foveae zone. Treatment with QIAPI 1™ (U.S. Pat. No. 9, 918, 996) was started immediately at doses of three drops sublingually each two hours during day.

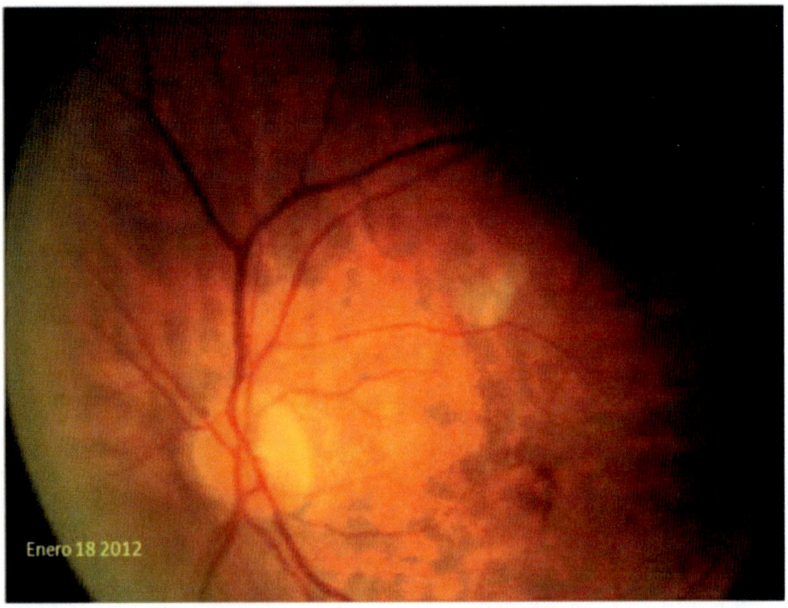

Fig. (13). Three and half months later, the anatomic improvement was very encouraging. There are no hemorrhages or exudates. The visual acuity was 20/50.

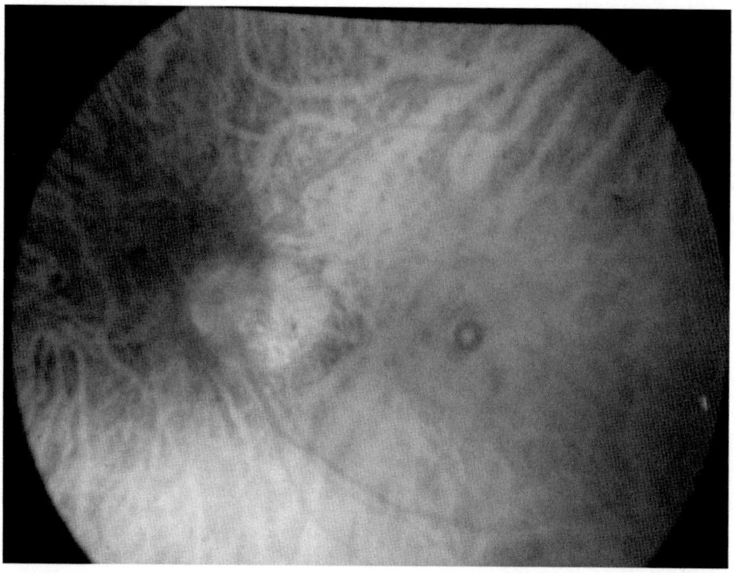

Fig. (14). Photograph taken with infrared light, that allows a much better observation of the choroidal circulation. The rounded black formation is the hemorrhage and the subtle peripheral circle around is given by edematous tissue.

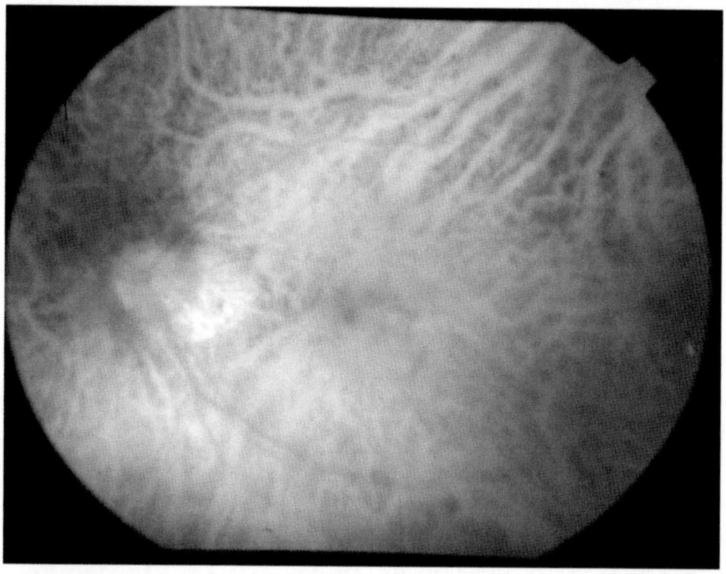

Fig. (15). After five weeks of treatment with QIAPI 1™ (U.S. Pat. No. 9, 918, 996), at doses of three drops sublingually each two hours, there was a good improvement in the infrared photographic appearance, the rounded black dot, given by the hemorrhage disappeared and the subtle circle now became smaller than the previous photograph; so the edematous tissue diminished significantly. The integrity of the choroidal vascular layer can be noticed. The final vision of this patient was 20/25.

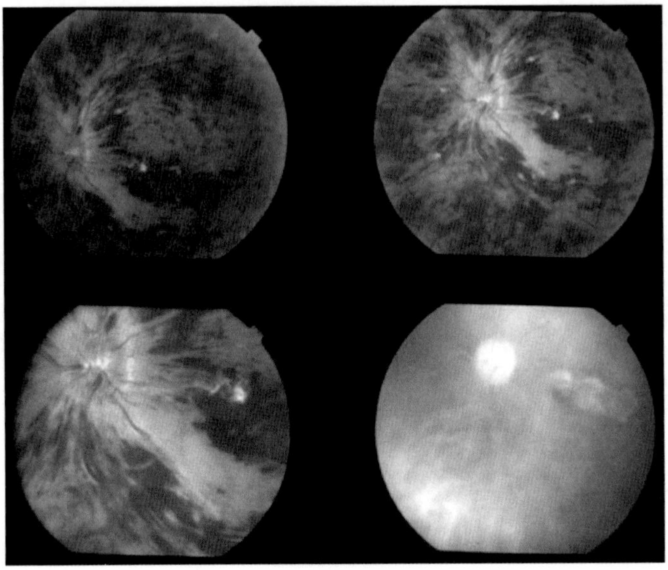

Fig. (16). The photograph shows multiple hemorrhages and edema of the male patient, 44 years old; with sudden onset of two weeks of evolution characterized by vision loss. The general examination was unremarkable. The observation of the choroid with monochromatic light of 800 nm of wavelength (white star) showed a very affected choroidal circulation. Treatment with QIAPINE™ (U.S. Pat. No. 9, 918, 996) at doses of three drops sublingually each hour was started the same day.

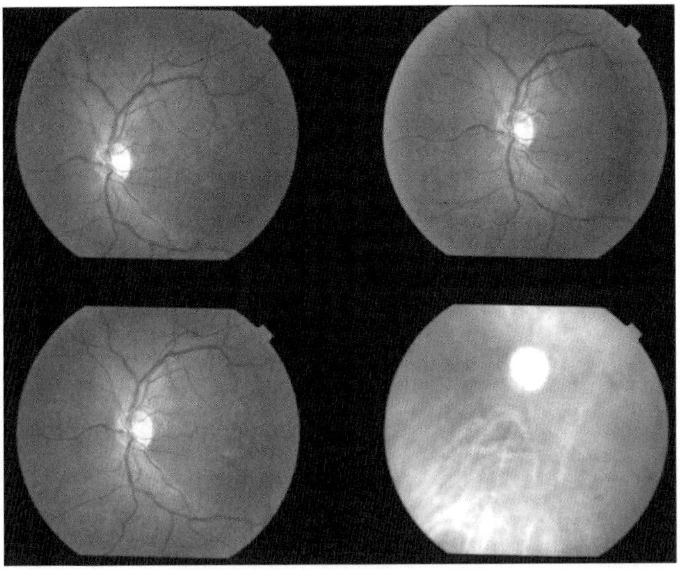

Fig. (17). Four months later the clinical picture shows an astonishing improvement, both retinal and choroidal (white star) circulation showed a very good improvement in the anatomical characteristics and the function was also recovered markedly, final vision 20/60; the only employed treatment was QIAPINE™ (U.S. Pat. No. 9, 918, 996), three drops each hour sublingually during day time for several months.

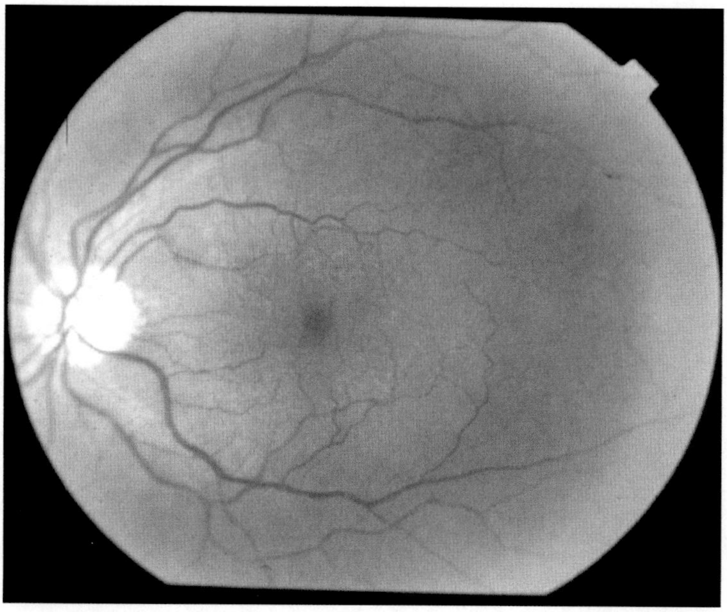

Fig. (18). A female patient, 56 years old, with complaints of subtle vision loss in the left eye of few months of evolution. At examination, the macular area showed compatible changes with primary premacular gliosis. The vision of the left eye at first examination was: 20/50.

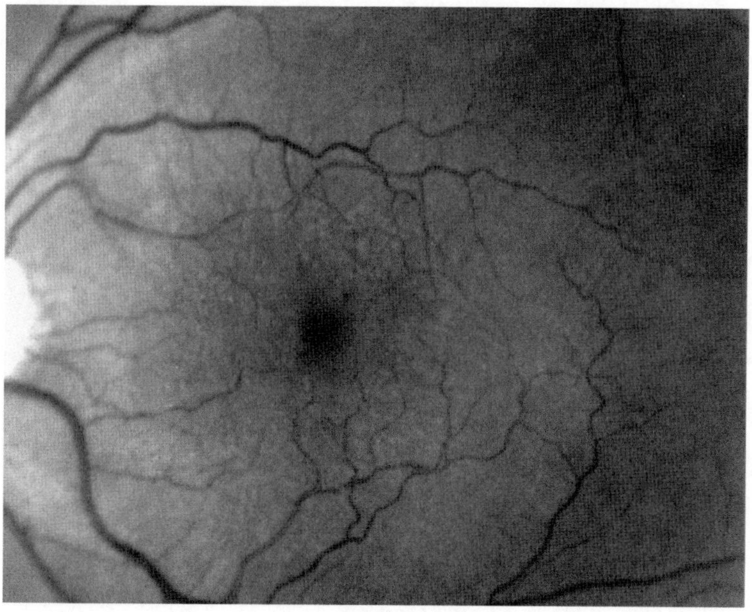

Fig. (19). Enlargement of the same patient of Fig. (**31**). Macular anatomy was moderately distorted. Usually this distortion is progressive and the vision falls markedly.

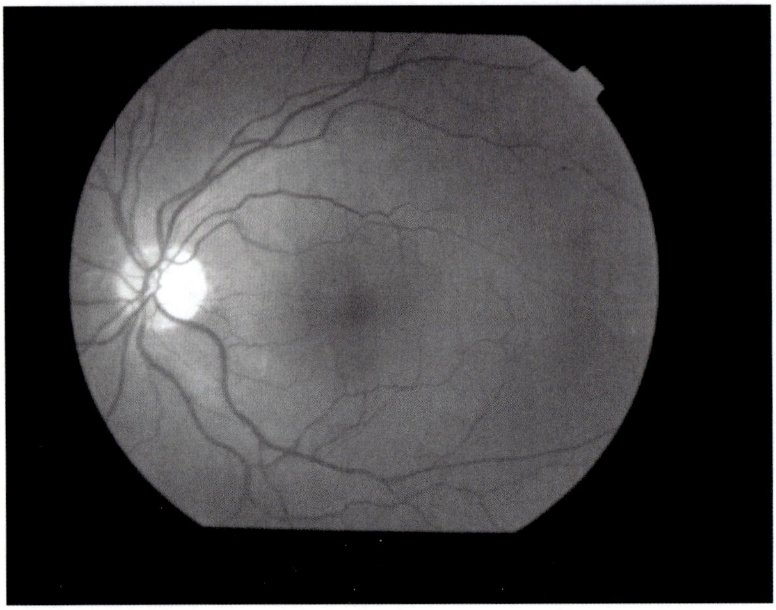

Fig. (20). At the sixth month of treatment. Photograph shows the improvement of the smoothness at retinal surface and hence the recovery of visual acuity to 20/25.

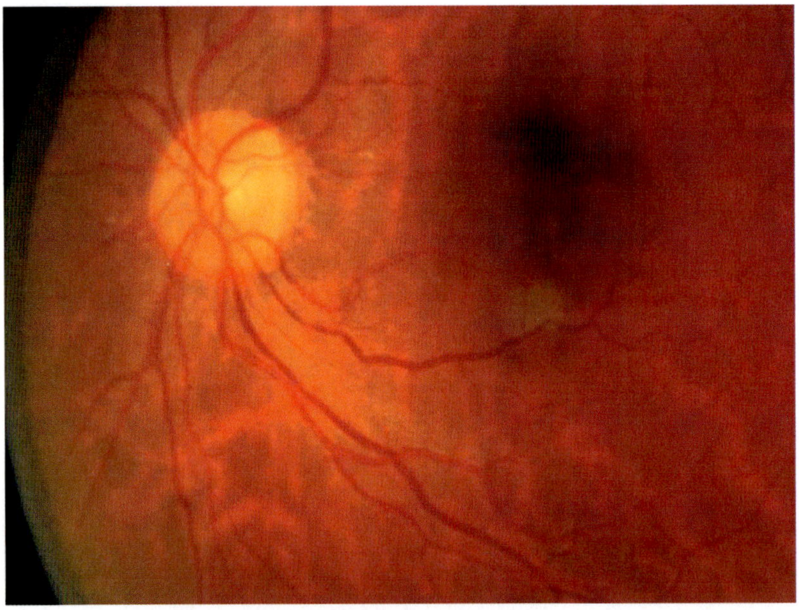

Fig. (21). Photograph in color of the patient one year later. This patient also was treated with QIAPI 1® (U.S. Pat. No. 9, 918, 996), three drops sublingually every two hours during day time.

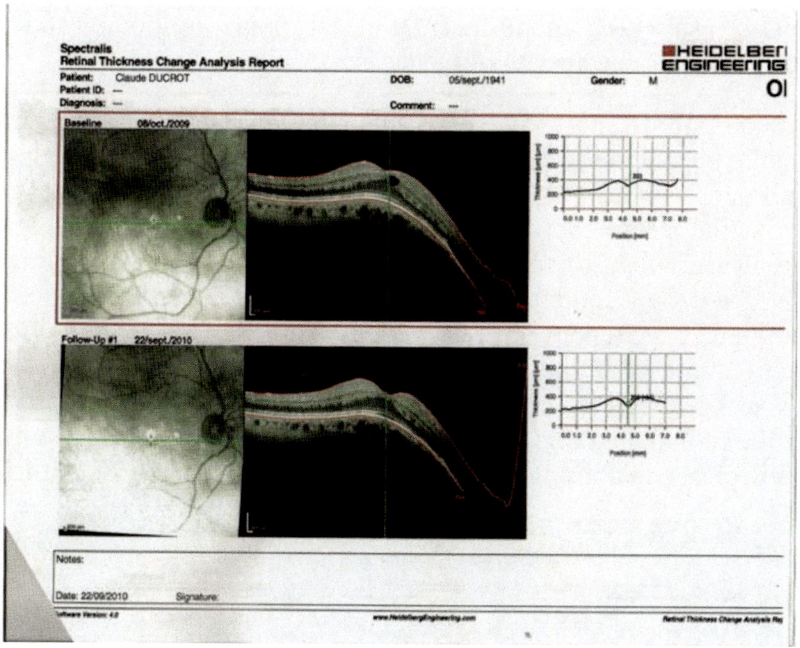

Fig. (22). Comparative Retinal thickness analysis of female patient diagnosed with Age Related Macular Degeneration. The OCT study taken in October 2009 (upper side) shows an intra-retinal cystic space. In the OCT study conducted in September 2010, macular regions showed a dramatic recovery with QIAPI 1® (U.S. Pat. No. 9, 918, 996), three drops sublingually every two hours during day time. The patient lives in France.

Improvement in anatomic and functional alterations of macular was surprising when generation and distribution of chemical energy was intensified. Many processes, known or unknown; that they involved so the retina can keep the form and function, depends on the chemical energy that emanates from the melanin; as any chemical reaction involves exchange of energy.

Monoclonal Antibodies Anti-VEGF in the Treatment of ARMD

The use of specific monoclonal antibodies that inactivate VEGF, produced by the body naturally, eye is a very important peptide in the generation of new blood vessels (angiogenesis). It was initially thought that using very small anti-VEGF antibody and the injected dose as close as possible to the place where you want to combat neo-vascularization, systemic problems are expected, given that VEGF is a compound that is normally found in the human body; they could be avoided to a large extent; but not, as in experiments with tag antibodies, anti-VEGF appears in the entire system (liver, brain, heart, intestines, lung, *etc.*) and a quarter of the patients treated with intraocular injections have systemic complications that merit hospitalization.

Unfortunately, it has been found that the effect of the antibodies anti-VEGF is very fleeting, favoring the growth of metastases.

The recurrence rate of choroidal neovascular membranes that complicates age-related macular degeneration within 5 years may be as high as 50% and frequently these membranes extend into the foveal avascular zone [15]. To evaluate the results of existing treatments for AMD, the primary outcome variable is visual acuity. The measurement of contrast sensitivity and reading speed for large print text was affected by supplement visual acuity. The electroretinogram (ERG) which is defined as the record of the electrical activity of the retina is an ancillary component of studies that may be performed in patients affected by AMD. When the studies using Electroretinogram are conducted, the response is generated entirely by the retinal area under stimulation. The pattern-reversal ERG showed a progressive decrease in amplitude with increased artificial central scotoma size.

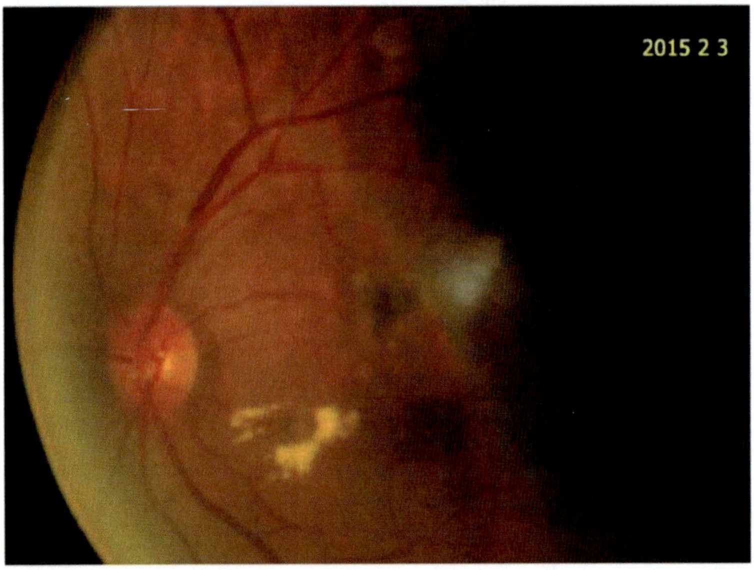

Fig. (24). Female patient of 55 years of age, no history of importance for the current condition; who started with progressive distortion of the central vision of the OI, attended consultation 4 months that started experiencing the problem. OI central vision was 20/80 with distortion. As shown in the photographs, two zones with pigmentary changes and the lower lesion showed important exudates around it. The macula can be seen preserved, but affected by the swelling of tissues. Medical treatment was initiated to enhance the water molecule dissociation.

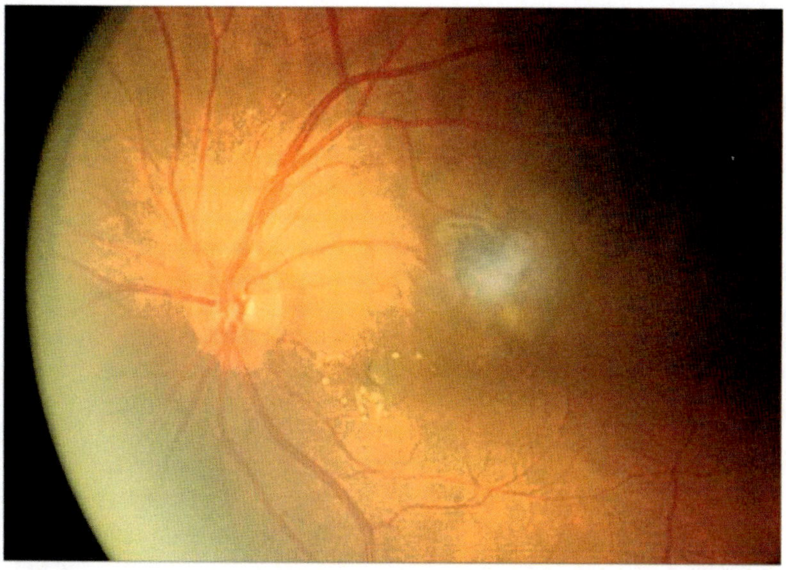

Fig. (25). October 20, 2015. After six months of treatment, appearance of lesions is better noticeable; amount of exudates diminished considerably, this in inferior lesion. The pigmented changes in the upper part of macula showed a significant improvement.

Bevacizumab (Avastin, Genentech) is the most commonly used drug in the USA for the treatment of neovascular AMD, despite the absence of large scale clinical-trial data supporting its use [16]. It is a recombinant humanized monoclonal immunoglobulin antibody, an anti-human vascular endothelial growth factor (VEGF) agent that has received approval as an adjunct treatment for colorectal cancer [17]. Intra vitreal injections of Bevacizumab for off-label use have demonstrated favorable short term results in the eyes with macular edema secondary to neovascular age-related macular degeneration (AMD), diabetic retinopathy, and central or branch retinal vein occlusion (CRVO, BRVO). However, in spite of the expanded application of Bevacizumab worldwide, the pharmacokinetic profile of this anti VEGF agent after intra vitreal injection in humans has not been determined clearly. The molecular weight of Bevacizumab is 149 Kd.

The available data are scarce and incomplete and most of the data are obtained from the manufacturer. With respect to the half-life independent studies, in human non-vitrectomized eyes, the aqueous half-life of 1.5 mg intravitreally injected bevacizumab is 9.82 days (Krohne and Eter, 2008). It is thought that the half-life in the vitreous body is similar to that of aqueous humor, but not in vitrectomized eyes, because the half-life of the substances studied (*e.g.* triamcinolone acetonide) is significantly shortened. And the studies that reported the loss of clinical effect of bevacizumab in the eyes with diabetic macular edema vitrectomized, theory postulated to be due to the short half-life of anti-VEGF agent due to vitreous absence [18]. However, the finding of fragments of VGEF membrane naïve receptors in the vitreous of eyes treated with intraocular injections of anti-VEGF agents, as well as the gradual loss of effect despite repeated injections, suggest that intraocular injections of Avastin and Lucentis induce an immune response specifically against membrane receptor of VEGF on endothelial cells of the retina and choroid mainly, so the effect of these injections tends to perpetuate due to repeated immunizations.

Anti VEGF injections induce a sustained immune response against VEGF receptor of the vascular endothelial cells of the eye and contribute to increase VEGF concentrations.

The explanation is that the organism initiates an immune response against injected intraocular monoclonal antibodies, and the native VEGF membrane receptors are the image in mirror of antibodies anti - VEGF, exogenous so it has been observed that the body destroys their own receptors VEGF, given the huge similarity with monoclonal antibodies, which also explains that around 25% of patients treated with antibodies, monoclonal, presented systemic vascular events that valued hospitalization. The indiscriminate use of anti-VEGF antibody induces sensitization of long duration against native VEGF membrane receptors, which results in a very important systemic involvement, since native VEGF receptors are indispensable to life.

In experimental animals, the drugs have been identified in the choroid, retina, ciliary body, and blood plasma. The injection in one eye reached the other. That anti VEGF reached the choroid which was expected since that is where they must act to influence the sub retinal neovascularization.

In 2005, clinical trials were conducted to study the effect of ranibizumab in patients with macular degeneration related to age under an as-needed regimen. In the reported studies, 98% of patients were phototype 1-3, *i.e.* white skin, blonde hair, blue eyes, green or gray.

ERG is an objective and reliable assessment for the retina function *in vivo* [19]. Two traditional rod pathways are known to exist in mammals. The primary pathway for rod signals is transmission from rods →rod bipolar cells → all amacrine cells → cone ON and OFF bipolar cells → ganglion cells. The second pathway for rod signals is from rods → cones (through gap junctions) → ON and OFF cone bipolar cells V ganglion cells.

Although mfERG responses after injection improved compared with the baseline, the amplitude and implicit time were in the sub-normal range. Analysis showed that intravitreal bevacizumab injection contributed to the improvement of retinal function, but did not result in complete recovery [20].

Data from psychophysical tests and the results from electrophysiological examinations indicated a global reduction of retinal function in ARMD which seemed to be present not only in the macula but also elsewhere in the retina [21].

Vascular endothelial growth factor (VEGF) is a key regulator of physiological angiogenesis (Vasculogenesis) during embryogenesis, skeletal growth and reproductive functions [22]. VEGF is also implicated in pathological angiogenesis as in tumors, intraocular disorders and other conditions. The biological effects of VEGF are mediated by two receptor tyrosine kinases (RTKs in humans), VEGFR-1 and VEGFR-2, which differ considerably in signaling properties.

Among growth factors, vascular endothelial growth factor (VEGF) has emerged as a prime mediator of endothelial cell survival and function. VEGFs are a family of polypeptides with diverse functions. The vital importance of VEGF in integrity of vasculature is underscored by the observation that absence of a single allele of VEGF is embryonic lethal due to severe derangement in endothelial cell development [23].

Furthermore, for the genesis of endothelium, organization of endothelial cells into vascular beds is regulated by VEGF and its counterparts VEGF-R1 and VEGF-R2. VEGF is also called vascular permeability factor as it regulates passage of fluid across endothelial cell layer.

VEGF and VEGF receptors regulate important metabolic functions of the cell such as protein synthesis and other metabolic activities which have not yet been identified. At first, VEGF increases tyrosine phosphorylation of VEGF-R1 and VEGF-R2, which indicates activation of the receptors. Also, VEGF increases insulin receptor substrate-1 (IRS-1) tyrosine phosphorylation and its association with VEGF-R2, as it has been linked to most VEGF-related effects in endothelial cells. Among other effects reported of VEGF it has been observed that phosphorylation of endothelial nitric oxide synthase (eNOS) thereby stimulates nitric oxide generation.

Several vascular endothelial growth factors (VEGF)-targeted agents, administered either as single agents or in combination with chemotherapy, have been shown to produce a short term benefit in patients with advanced-stage malignancies. VEGF-targeted therapies were initially developed with the notion that they would inhibit new blood vessels growth and thus starve tumors of necessary oxygen and nutrients. It has become increasingly apparent, that the therapeutic benefit associated with VEGF-targeted therapy is complex, and involves multiple mechanisms [24].

In mouse, after application of an anti-VEGFR-2 antibody, there is an increase in plasma VEGF levels that was induced in a dose-dependent manner.

A major obstacle compromising successful application of angiogenesis inhibitor is the empiricism associated with determining an effective biological therapeutic dose because these drugs express optimum therapeutic activity below the maximum tolerated dose, if such a dose could be defined [25]. VEGF family consists of VEGF-A, also called VEGF, which is produced in a number of splice variant isoforms, especially the VEGF 121 and VEGF 165. Many tissues with mitosis disordered (*e.g.* tumors) express abundant levels of VEGF. Environmental, genetics and epigenetics factor, especially hypoxia can lead to VEGF induction or up-regulation in tissues.

VEGF receptor (VEGFR)-2, also called KDR in humans, appears more functionally important of the two and can mediate signaling events involved in endothelial cell mitogenesis, migration, survival and vascular permeability.

Anti VEGF regimen does not improve vision over the long term (Samir 2013). All anti- VEGF agents have shown promising results with regard to the regression of neovascularization, but they are limited by their short duration [26]. The local adverse effects described are tractional retinal detachment, vitreous hemorrhage; and extreme fluctuations in intraocular pressure. The anti- VEGF intravitreal or systemic injection reduces unbound VEGF, this is biologically active; however, this seems to result in the raise of Connective Tissue Growth Factor (CTGF) and thus promotes a switch from angiogenesis to fibrosis. Other also serious ocular side effects are endophthalmitis, cataract, macular hole due to vitreous traction; *etc.*

Systemic adverse effects have been reported: strokes, myocardial infarctions, acute elevation of blood pressure, kidney failure, especially in diabetic patients, worsening of metastasis; arteries or venous -thrombotic events.

Examples of Clinical Cases Fig. (26 - 31)

The photographs below show the complications that we have observed in patients that have been injected into other places, monoclonal antibodies are not considered.

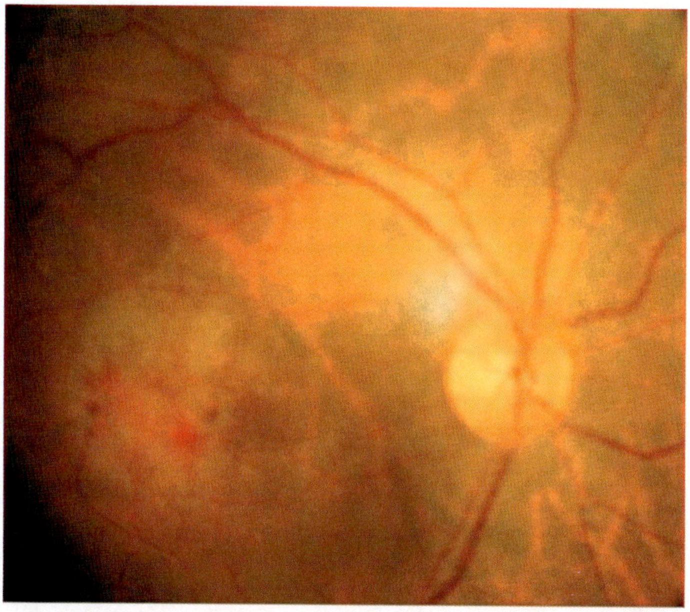

Fig. (26). Intraocular injections of anti-VEGF antibody produced repeated bleeding, edema, explainable by progressive Endothelial dysfunction secondary to immune response inducing iatrogenic form against natives VEGF receptors of endothelial cells.

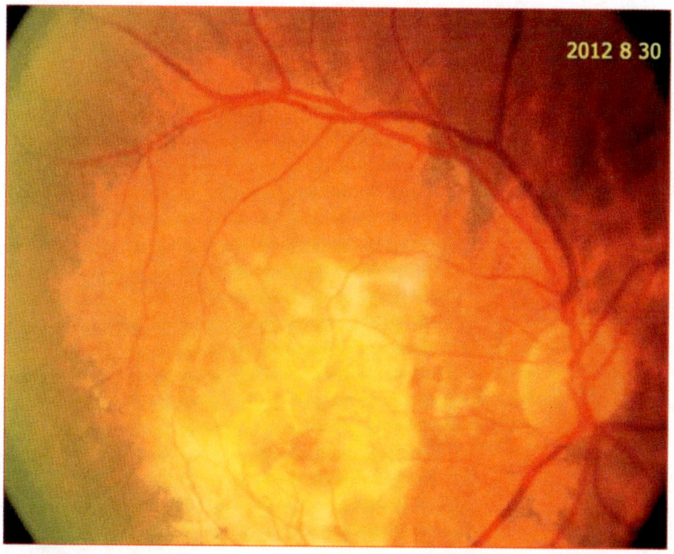

Fig. (27). The scars that occurred in the macular area after intraocular injections of monoclonal antibodies anti-VEGF were different from the scars that are observed during the natural course of the disease. The constant presence of bleeding of different sizes was usual.

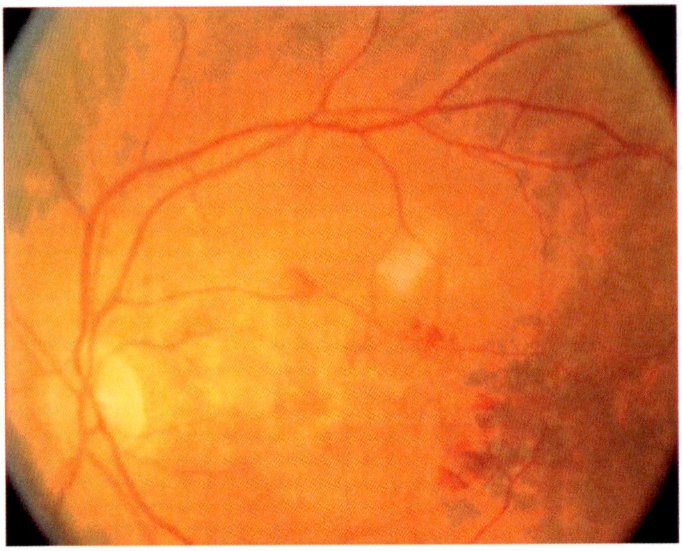

Fig. (28). A moderate distortion of the view, and when treated with intra ocular application of monoclonal antibodies, anti-VEGF presented a very fleeting improvement followed by a progressive deterioration due to vascular endothelial dysfunction. Bleeding from capillaries of neoformation or normal vessels that altered anatomically, suggesting that the endothelial VEGF receptors are a normal component of blood vessels and allow regulation, molding, and modulation of vascular functions elementary and indispensable for the functioning of the tissue which can be seen in the picture. When these VEGF endothelial receptors were affected, the tissue exhibited functional and anatomical alterations that were progressive and severely affected its function, not only in the retina, but in other parts of the body (brain, heart, lung; *etc.*).

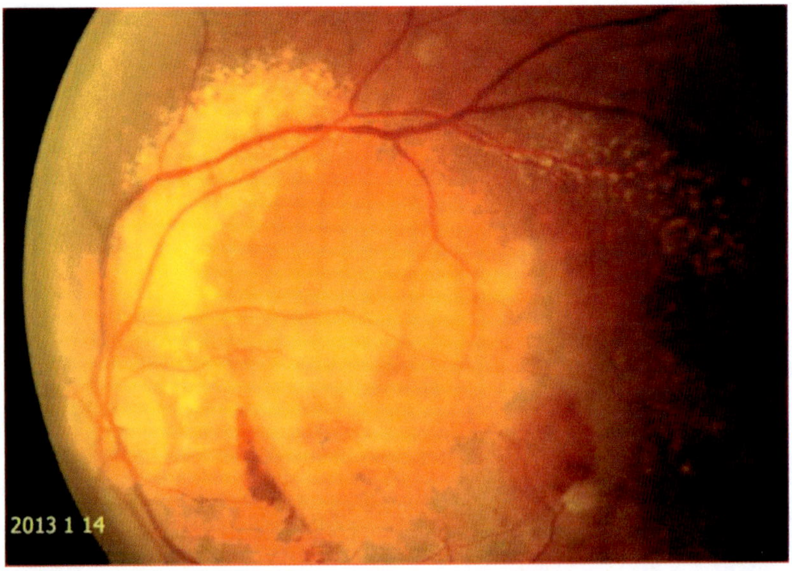

Fig. (29). Photograph was taken a few weeks later and belongs to the same patient from the previous picture. However, the application of anti-VEGF monoclonal antibodies increased hemorrhages, and the swelling was more intense. It is likely that the problem was in the "immunization" *versus* naïve VEGF receptors that was progressive with repeated injections and persisted for years.

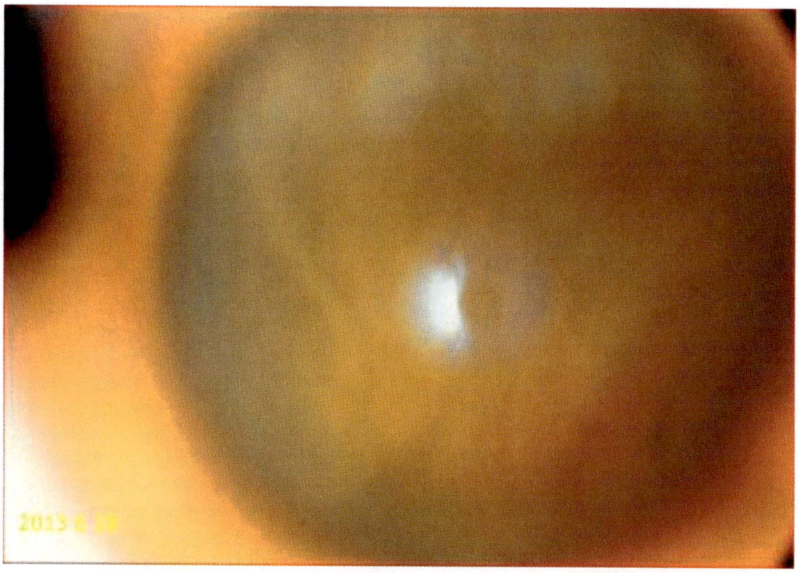

Fig. (30). Alterations were progressive and seriously undertook the complex and different mechanisms through which blood vessels are regulated as well as their participation in processes such as healing. The picture shows a detached retina that despite having been operated several times, developed a fibrous proliferation that made the retina detachment irreparable.

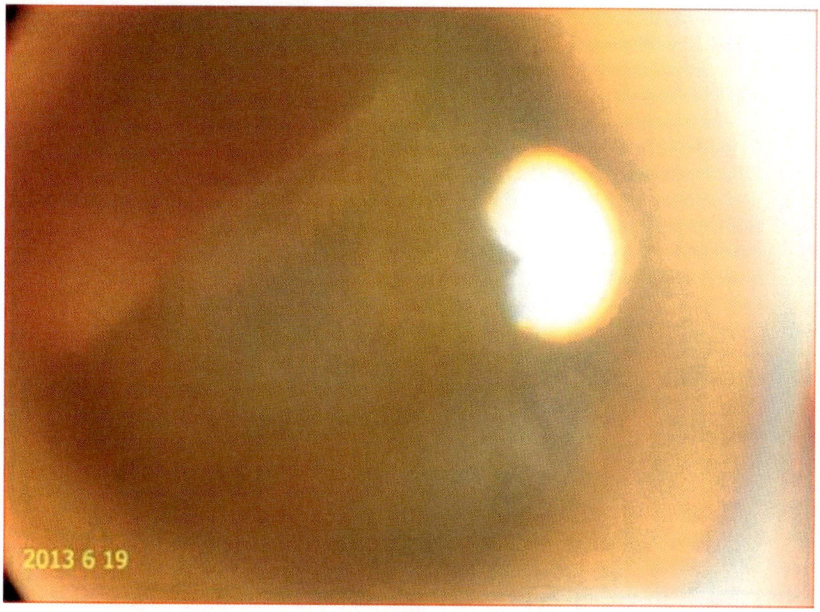

Fig. (31). This photograph allows us to observe a huge retinal tear at the top left of the picture, which increased due to alterations in the normal repair mechanisms secondary to intraocular injections of anti-

VEGF monoclonal antibodies.

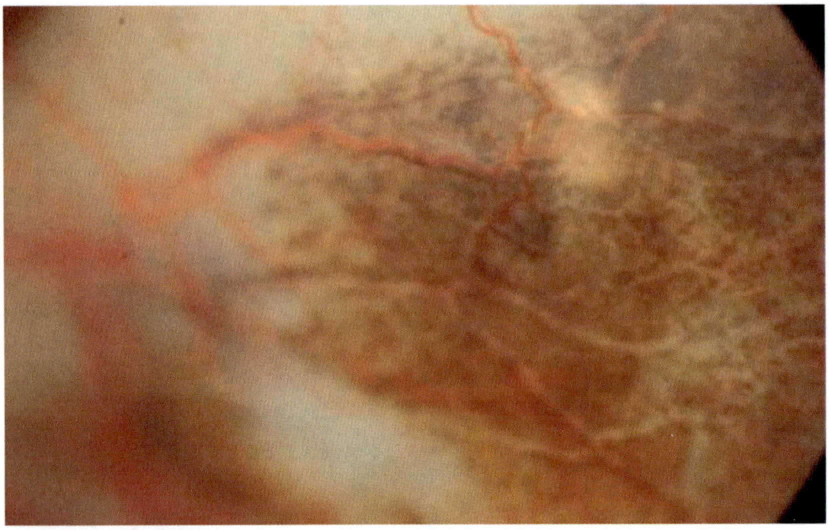

Fig. (32). Microphotograph of sclera´s nevus in a live patient. The conjunctiva blood vessels that lie over the sclera´s nevus (better focused in the photography) are fully permeable and filled with the blood stream. However, the blood vessels placed inside or closest to the sclera´s nevus are closed. This is an indirect proof that the oxygen diatomic levels nearest melanin are higher than at normal sclera tissue. The explanation of the closed vessels is that the elevated levels of diatomic oxygen coming from melanin´s water dissociation have a powerful anti-angiogenic effect.

Vitamins, Supplements and Anti-oxidants in the Treatment of ARMD

Whether vitamins, antioxidant, and supplements are beneficial or harmful is uncertain [27]. Many people take vitamins, antioxidant, and supplements, believing that this will improve their health and prevent diseases. Many primary or secondary prevention trials of anti-oxidant supplements have been conducted to prevent several diseases. In a Systematic Review and Meta-analysis, Bjelakovic and Nikolova (2007) found that antioxidant supplements increase all-cause mortality.

This is congruous with recommendations of National Eye Institute, USA (below):

"Those who strictly write that a treatment preventing the development of macular degeneration related to age is not known", referring to the ARED study that sought to establish the benefit of vitamins and minerals in the therapeutic management of degeneration macular age-related [28].

CONCLUSION

The pathogenesis of the three leading causes of blindness in the world (macular degeneration related to age, diabetic retinopathy and glaucoma) is likely to be reconsidered in the light of the discovery of unsuspected intrinsic property of melanin to transform light energy into chemical energy by means of the dissociation of the molecule of water, such as chlorophyll in plants.

In the vast majority of published studies, the pigment is usually of secondary importance, not focused on pigmentary alterations; This is understandable given that melanin is considered since 1820 as a simple sunscreen that protects humans from dangerous UV rays, and in the case of the eyeball, it has been observed that it absorbs excess light, thus allowing better image quality.

It can be said that alterations of pigment are at the beginning and not at the end, since the generation and distribution of chemical energy are crucial aspects in the function of cells and therefore, in the function of the organism. In order to discover the main causes of amaurosis, there is a possibility of finding new treatments, more effective and with fewer side effects for major vision problems which are common throughout the world.

HUMAN AND ANIMAL RIGHTS

No humans were used in this research. All animal research procedures followed were in accordance with the standards set forth in the eighth edition of Guide for the Care and Use of Laboratory Animals published by the National Academy of Sciences, The National Academies Press, Washington, D.C.

CONFLICT OF INTEREST

The author (editor) declares no conflict of interest, financial or otherwise.

ACKNOWLEDGEMENTS

We thank the continuous financial support of the Center for Studies of Human Photosynthesis® for the realization of this studio.

REFERENCES

[1] Tomany, S.C.; Wang, J.J.; Van Leeuwen, R.; Klein, R.; Mitchell, P.; Vingerling, J.R.; Klein, B.E.; Smith, W.; De Jong, P.T. Risk factors for incident age-related macular degeneration: pooled findings from 3 continents. *Ophthalmology,* **2004**, *111*(7), 1280-1287. [http://dx.doi.org/10.1016/j.ophtha.2003.11.010] [PMID: 15234127]

[2] Dasari, B; Prasanthi, JR; Marwarha, G; Singh, BB; Ghribi, O Cholesterol-enriched diet causes age-related macular degeneration-like pathology in rabbit retina. *BMC Ophthalmol.,* **2011**, *11*, 22. [http://dx.doi.org/10.1186/1471-2415-11-22]

[3] Ding, J.D.; Lin, J.; Mace, B.E.; Herrmann, R.; Sullivan, P.; Bowes Rickman, C. Targeting age-related macular degeneration with Alzheimer's disease based immunotherapies: anti-amyloid-β antibody

attenuates pathologies in an age-related macular degeneration mouse model. *Vision Res.,* **2008**, *48*(3), 339-345.
[http://dx.doi.org/10.1016/j.visres.2007.07.025] [PMID: 17888483]

[4] Javitt, J.C.; Zhou, Z.; Maguire, M.G.; Fine, S.L.; Willke, R.J. Incidence of exudative age-related macular degeneration among elderly Americans. *Ophthalmology,* **2003**, *110*(8), 1534-1539.
[http://dx.doi.org/10.1016/S0161-6420(03)00495-0] [PMID: 12917168]

[5] Khandhadia, S.; Cherry, J.; Lotery, A.J. Age-related macular degeneration. *Adv. Exp. Med. Biol.,* **2012**, *724*, 15-36.
[http://dx.doi.org/10.1007/978-1-4614-0653-2_2] [PMID: 22411231]

[6] Ozkiris, A. Anti-VEGF agents for age-related macular degeneration. *Expert Opin Ther Pat.,* **2010**, *20*(1), 103-118.
[http://dx.doi.org/10.1517/13543770902762885]

[7] Ohno-Matsui, K. Parallel findings in age-related macular degeneration and Alzheimer's disease. *Prog. Retin. Eye Res.,* **2011**, *30*(4), 217-238.
[http://dx.doi.org/10.1016/j.preteyeres.2011.02.004] [PMID: 21440663]

[8] Grammas, P.; Sanchez, A.; Tripathy, D.; Luo, E.; Martinez, J. Vascular signaling abnormalities in Alzheimer disease. *Cleve. Clin. J. Med.,* **2011**, *78* (Suppl. 1), S50-S53.
[http://dx.doi.org/10.3949/ccjm.78.s1.09] [PMID: 21972332]

[9] Flamendorf, J.; Agrón, E.; Wong, W.T.; Thompson, D.; Wiley, H.E.; Doss, E.L.; Al-Holou, S.; Ferris, F.L., III; Chew, E.Y.; Cukras, C. Impairments in dark adaptation are associated with age-related macular degeneration severity and reticular pseudodrusen. *Ophthalmology,* **2015**, *122*(10), 2053-2062.
[http://dx.doi.org/10.1016/j.ophtha.2015.06.023] [PMID: 26253372]

[10] Dentchev, T.; Milam, A.H.; Lee, V.M.; Trojanowski, J.Q.; Dunaief, J.L. Amyloid-beta is found in drusen from some age-related macular degeneration retinas, but not in drusen from normal retinas. *Mol. Vis.,* **2003**, *9*, 184-190.
[PMID: 12764254]

[11] Gregori, G.; Wang, F.; Rosenfeld, P.J.; Yehoshua, Z.; Gregori, N.Z.; Lujan, B.J.; Puliafito, C.A.; Feuer, W.J. Spectral domain optical coherence tomography imaging of drusen in nonexudative age-related macular degeneration. *Ophthalmology,* **2011**, *118*(7), 1373-1379.
[PMID: 21388687]

[12] Solis Herrera, A.; Arias Esparza, M. de C.; Solis Arias Ruth, I.; Solis Arias Paola, E.; Solis Arias Martha, P. The unexpected capacity of melanin to dissociate the water molecule fills the gap between the life before and after ATP. *Biomed. Res.,* **2010**, *21*(2), 221-225.

[13] Kaarniranta, K.; Salminen, A.; Haapasalo, A.; Soininen, H.; Hiltunen, M. Age-related macular degeneration (AMD): Alzheimer's disease in the eye? *J. Alzheimers Dis.,* **2011**, *24*(4), 615-631.
[PMID: 21297256]

[14] Ding, J.D.; Lin, J.; Mace, B.E.; Herrmann, R.; Sullivan, P.; Bowes Rickman, C. Targeting age-related macular degeneration with Alzheimer's disease based immunotherapies: anti-amyloid-β antibody attenuates pathologies in an age-related macular degeneration mouse model. *Vis. Res.,* **2008**, *48*(3), 339-345.
[http://dx.doi.org/10.1016/j.visres.2007.07.025] [PMID: 17888483]

[15] Birch, D.G.; Anderson, J.L.; Fish, G.E.; Jost, B.F. Pattern-reversla Electroretinographic Acuity in Untreated Eyes with subfoveal neovascular membranes. *Invest. Ophthalmol. Vis. Sci.,* **1992**, *33*, 1097-15.

[16] Martin, D.F.; Maguire, M.G.; Ying, G.S.; Grunwald, J.E.; Fine, S.L.; Jaffe, G.J. Ranibizumab and bevacizumab for neovascular age-related macular degeneration. *N. Engl. J. Med.,* **2011**, *364*(20), 1897-1908.
[http://dx.doi.org/10.1056/NEJMoa1102673] [PMID: 21526923]

[17] Krohne Tim, T.U.; Eter, N.; Holz, F.G.; Meyer, C.H. Intraocular pharmacokinetics of bevacizumab after a single intravitreal injection in humans. *Am. J. Ophthalmol.*, **2008**, *146*(4), 508-512. [http://dx.doi.org/10.1016/j.ajo.2008.05.036] [PMID: 18635152]

[18] Yanyali, A.; Aytug, B.; Horozoglu, F.; Nohutcu, A.F. Bevacizumab (Avastin) for diabetic macular edema in previously vitrectomized eyes. *Am. J. Ophthalmol.*, **2007**, *144*(1), 124-126. [http://dx.doi.org/10.1016/j.ajo.2007.02.048] [PMID: 17601433]

[19] Lei, B. Rod-driven OFF pathway responses in the distal retina: dark-adapted flicker electroretinogram in mouse. *PLoS One,* **2012**, *7*(8), e43856. [http://dx.doi.org/10.1371/journal.pone.0043856] [PMID: 22937111]

[20] Park, J.Y.; Kim, S.H.; Park, T.K.; Ohn, Y.H. Multifocal electroretinogram findings after intravitreal bevacizumab injection in choroidal neovascularization of age-related macular degeneration. *Korean J. Ophthalmol.*, **2011**, *25*(3), 161-165. [http://dx.doi.org/10.3341/kjo.2011.25.3.161] [PMID: 21655040]

[21] Walter, P.; Widder, R.A.; Lüke, C.; Königsfeld, P.; Brunner, R. Electrophysiological abnormalities in age-related macular degeneration. *Graefes Arch. Clin. Exp. Ophthalmol.*, **1999**, *237*(12), 962-968. [http://dx.doi.org/10.1007/s004170050331] [PMID: 10654164]

[22] Ferrara, N.; Gerber, H-P.; LeCouter, J. The biology of VEGF and its receptors. *Nat. Med.*, **2003**, *9*(6), 669-676. [http://dx.doi.org/10.1038/nm0603-669] [PMID: 12778165]

[23] Denis, F.; Chen, X.; Nese, A. Choudhury G.G.; Madaio M., VEGF regulation of endothelial nictirc oxide synthase in glomerular endothelial cells. *Kidn. Int.*, **2005**, *68*, 1648-1659. [http://dx.doi.org/10.1111/j.1523-1755.2005.00575.x] [PMID: 16164642]

[24] Denis, F.; Chen, X.; Nese, A. Choudhury, G.G., Michael, M. VEGF regulation of endothelial nictirc oxide synthase in glomerular endothelial cells. *Kidney Int.*, **2005**, *68*, 1648-1659. [http://dx.doi.org/10.1111/j.1523-1755.2005.00575.x] [PMID: 16164642]

[25] Bocci, G.; Man, S.; Green, S.K.; Francia, G.; Ebos, J.M.; du Manoir, J.M.; Weinerman, A.; Emmenegger, U.; Ma, L.; Thorpe, P.; Davidoff, A.; Huber, J.; Hicklin, D.J.; Kerbel, R.S. Increased plasma vascular endothelial growth factor (VEGF) as a surrogate marker for optimal therapeutic dosing of VEGF receptor-2 monoclonal antibodies. *Cancer Res.*, **2004**, *64*(18), 6616-6625. [http://dx.doi.org/10.1158/0008-5472.CAN-04-0401] [PMID: 15374976]

[26] Osaadon, P.; Fagan, X.J.; Lifshitz, T.; Levy, J. A review of anti-VEGF agents for proliferative diabetic retinopathy. *Eye (Lond.),* **2014**, *28*(5), 510-520. [http://dx.doi.org/10.1038/eye.2014.13] [PMID: 24525867]

[27] Bjelakovic, N.; Dimitrinka, L.G.; Simoneti, R.G.; Christian, G. Mortality in randomized Trials of Antioxidant Supplements for Primary and Secondary Prevention. *JAMA,* **2007**, *297*(8) [http://dx.doi.org/10.1001/jama.297.8.842] [PMID: 17327526]

[28] The AREDS Formulation and Age-Related Macular Degeneration. https://nei.nih.gov/amd/summary

CHAPTER 8

Energy Production, the Key Role of the Melanin Molecule in the Human Body: Implications in the Context of Aging

Arturo Solís Herrera*, María del Carmen Arias Esparza, Ruth Isabel Solís Arias, Paola Eugenia Solís Arias and **Martha Patricia Solís Arias**

* *Human Photosynthesis™ Study Center, Avenida Aguascalientes Norte 607, Pulgas Pandas Sur, Aguascalientes, México; CP 20138*

Abstract: Nutrition has been defined as the study of food and nourishment, examining the nutritional content of different foods, the amount of nutrients required for healthy growth and function and how this varies for different people. The main focus was so far in carbohydrates, proteins, lipids, vitamins and minerals. The role of water, in spite to be the main constituent of the human body, was limited to volume in the circulatory apparatus, a simple solvent where metabolic reactions take place; an inner cleaner for the elimination of waste products, and maintenance of the body temperature and plasma volume.

Our finding, in the human eye initially, of the intrinsic property of melanin molecule to split and re-form the water molecule breaks the ground. Our body has the astonishing property to uses water as source of electrons, as in plants happen. This fact, previously unknown in human eukaryotic cell, represents a turning point in many fields of human knowledge, among them, the Nutrition concept.

Water is the source of energy by excellence and meals are merely the source of biomass. With meals our body makes skin, nails, hair, muscle, blood, neuron cells, bone; *etc.* Glucose is out of discussion, the perfect building block, thereby our organism is able to makes even nucleic acids arising from $C_6H_{12}O_6$. However, energy, defined as any thing that is able to produce some kind of change, is taken from water through the dissociation and re-formation of the molecule.

Melanin is the equivalent to the human chlorophyll. Both molecules possess the intrinsic capacity to transform photonic energy into free chemical energy, susceptible to be used by eukaryotic cell.

* **Corresponding author Arturo Solís Herrera:** Human Photosynthesis™ Study Center, Av. Aguascalientes Norte 607, Pulgas Pandas Sur, Aguascalientes, México; C.P. 20138; Tel/Fax: 524499160048/524499150042; E-mail: comagua2000@yahoo.com

Keywords: Biomass, Energy, Hydrogen, Macula, Melanin, Photosynthesis, Water dissociation.

THE ORIGIN OF LIFE

For how long has men been trying to elucidate the origin of life? It is a doubt that has fascinated human mind since the beginning of times. Employed technics have evolved, being more and more sophisticated each time, however the question remains: How did life begin?

Current scientific theories lack suitability. All of them are merely partial explanations about some small topics that do not integrate the entire process as a whole. Allow me to explain the beginning of life in accordance with my finding of the intrinsic property of melanin to split and re-form the water molecule.

Now days, theoretically the sunrays can be collected only by plants and other organisms that possess chlorophyll. The sun-chain can be schematized (wrongly) as follows (Fig. **1**):

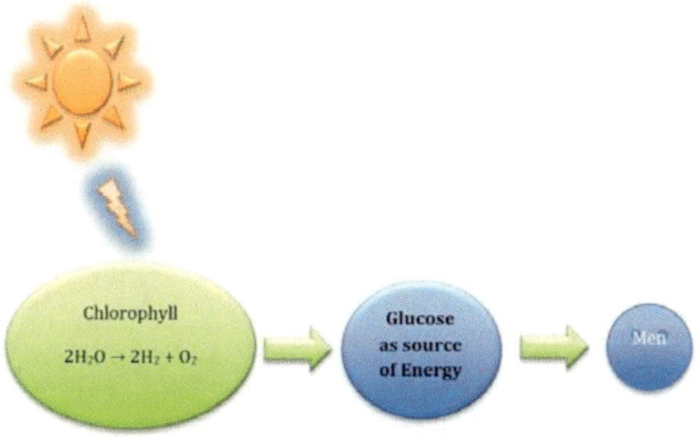

Fig. (1). Supposedly, chorophyll was the only molecule capable to transform light into chemical energy through water dissociation. Theoretically, the energy released is stored in chemical bounds of glucose. Thereafter, the energy of covalent bonds is released through the respiration of the cell.

The metabolic pathways implemented through the past decades trying to explain the metabolism of glucose as source of energy per excellence, are controversial in 95 %.

Astonishingly, the intrinsic property of melanin to split and re-form the water molecule changes dramatically the sun-chain- living things; because the very first spark of life is those that release free chemical energy, as follows (Fig. **2**):

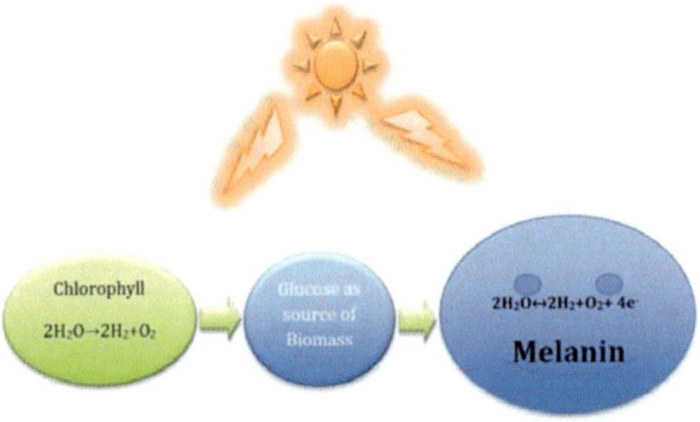

Fig. (2). With the discovery of the intrinsic property of melanin to dissociate the water molecule, now we have two molecules, widely disseminated in the biological kingdoms that can transform light into chemical energy through the dissociation of the water molecule: melanin and chlorophyll.

The beginning of life probably arises with transformation of photonic energy (from light) into free chemical energy (through water dissociation). However, cannot be explained yet just through chlorophyll.

The energy that comes from melanin is transported by molecular hydrogen (H_2) and high energy electrons (e^-) (Fig. **3**).

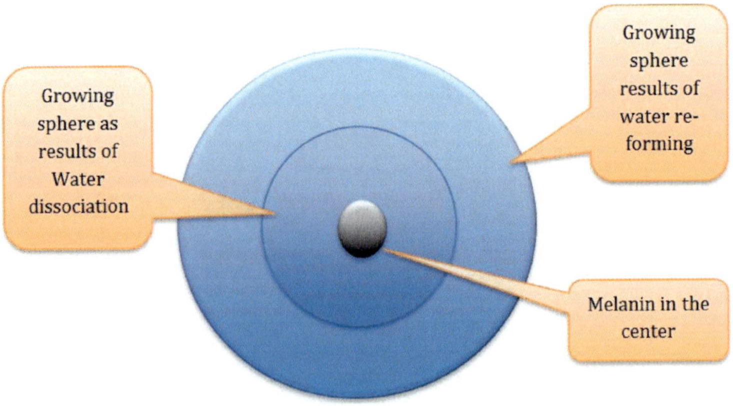

Fig. (3). The chemical energy is released by melanin symmetrically, spreading out in all directions, like growing spheres, that follow the rules of simple diffusion.

Glucose is the universal precursor of any organic matter in any living thing, but it cannot provide the energy that its own metabolism requires (Fig. **4**).

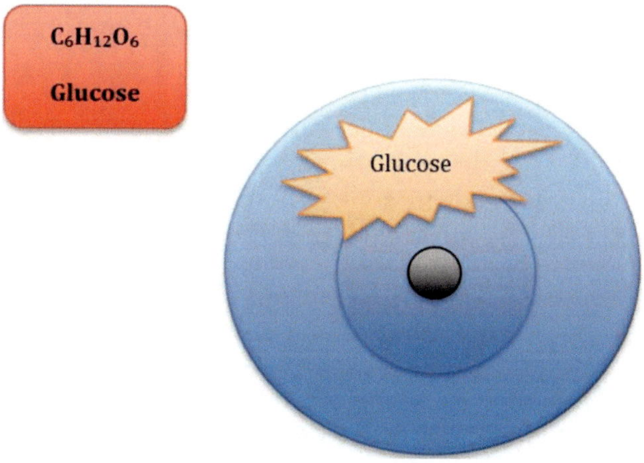

Fig. (4). Any organic molecule that fell into the domain of this growing spheres, changes its chemical behavior in a significative and highly ordered form.

The way in which melanin dissociates and re-form the water molecule, this is: liquid/gas/liquid/gas…. can be schematized as 0/1/0/1…. Like a binary code, that are abundant in the universe [1] (Fig. **5**).

Fig. (5). In the same way as the origin of times, every single molecule that takes part in some life process requires to be immersed in these growing spheres of free chemical energy in a constant form, night and day, through the life of the organism that contains it.

Cell Functions

There are basic cell functions essential for survival of the cell, and traditionally include the following:

1. - Obtaining food (nutrition) and oxygen (O_2) from the environment surrounding the cell.

Meals can be considered as a mixture of carbon chains of different lengths, branching, combination with nitrogen, hydrogen, oxygen; minerals, *etc.*, in different proportions among them and with a very assorted kind of bonds. We could say that carbon atom is the backbone of the biological molecules.

The replenishment of the biomass or the chemical composition of the body is the main aim of food. The most common atoms in our body are: Oxygen, carbon, hydrogen, and nitrogen, make up approximately 96 % of the total body chemistry.

These atoms and molecules formed by them as proteins, carbohydrates, lipids, polynucleotides, sometimes termed as molecules of life; are the inanimate raw ingredients from which all living things arise.

Our nutrition is based in the ingestion of relatively inanimate compounds that constitute the building blocks of our organism, an astonishingly our body imbued the life on them in minutes in some cases.

It is necessary to emphasize that even the minimum change in atoms or molecules that happen constantly into our body requires free chemical energy available. Therefore, energy is a very first unavoidable requirement in common at chemical level.

Food is merely source of building blocks for the buildup and replenishment of our biomass, it is important to emphasize that the energy content of meals is very low. Our body can store carbon chains, but cannot store energy.

Regarding oxygen (O_2) it is widely accepted its role as oxidizing agent over nutrient molecules for energy production, at the light of the unraveling of the Human Photosynthesis™ or the intrinsic property of melanin to split and re-form the water molecule, the role of oxygen over nutrient molecules is merely for biomass production and replenishment. Our body does not combine oxygen with glucose to get energy. Recall that in certain times were also widely accepted the earth was flat.

ATP, is Supposedly the Energy Currency of the Cell

Since the late 18th century, it had been known that sugar, burned as a fuel, behaves like a candle or coal. Carbon dioxide is formed when the carbon atoms of sugar, paraffin, or coal, unite with oxygen. And since 1940, was a growing idea that the cellular combustion of sugar captures much of the chemical energy of the sugar and stores it in ATP form, instead of wasting the energy as heath and light.

However, the questions remain: How does the ATP store and release energy? So far, the question still remains. How do the energy is used in the manufacture of even more complex chemicals, such as proteins and DNA? How do the chemical energy is converted to mechanical energy –movement-? Until today the chemo-osmotic theory of Michaels is accepted at most, however it still is only a theory.

In the yeast is perplexing how does the yeast cell grow and multiply on energy derived from sugar even in the absence of oxygen? 2 ATP molecules instead of 30 or 36 ATP molecules have not sense in any way from the energetic point of view, at light of Human Photosynthesis™, it is probably that the role ATP and thereby of mitochondria, is more oriented to the regulation of phosphate groups than energy production, recall that high levels of phosphate are toxic to the cell; thereby and possibly in yeast cells the phosphate requirements are less than in other organisms or species, especially in stages of the living thing when the consumption of phosphate groups is high, for instance during the buildup of cellular membrane, which is made of phospholipids, thereby, if the utilization of phosphate groups is elevated, then the necessity to the down-regulation of the process is less.

It is interesting our finding that ATP hydrolysis <u>absorbs</u> energy more than releasing it, and it can in inferred also from experiments from other researchers [2].

Mitochondria

The small amount of mitochondrial DNA codes for only 13 polypeptides in humans. These are mainly the hydrophobic cores of the major trans-membrane proton pumps, which are sticky and insoluble and difficult to move around the cell. All the remaining mitochondrial genes have migrated to the nucleus, and the hundreds of other mitochondrial proteins are now imported from the cytosol. Surprisingly, despite mitochondrial DNA is less well protected than nuclear DNA; significant mitochondrial variants are not observed in practice. In healthy people all the copies of the mitochondrial DNA are identical.

Mitochondria are subcellular organelles containing the Krebs cycle, fat oxidation pathway and the respiratory chain, which produces almost all the 70kg of ATP, used each day in the human body. They can be purified from tissue homogenates by differential centrifugation, using non-penetrant isotonic media to provide osmotic support. A few tissues (*e.g.* red blood cells, eye lens) have no mitochondria, and thereby cannot respire aerobically. In the case of red blood cells, the color is given by an iron porphyrin termed heme; by other side the magnesium porphyrin is the chlorophyll prosthetic group (see below). Furthermore, both molecules have several chemical similarities, but more important: both molecules can split irreversible the water molecule. Thereby the dissociation of the water molecule is probably the real source of energy of erythrocyte.

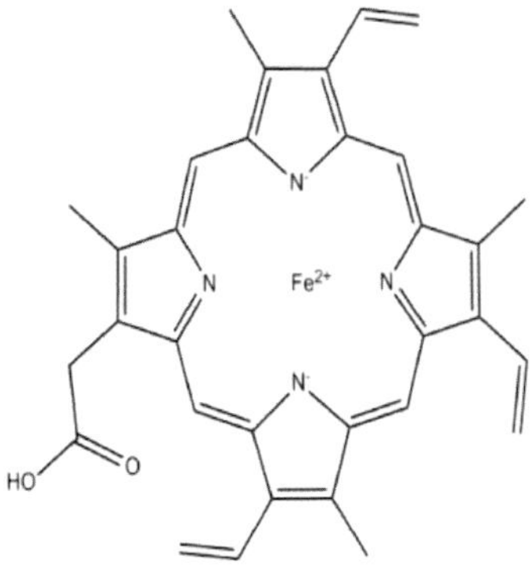

Heme group
(Hemoglobin)

Chemical Formula: $C_{32}H_{28}FeN_4O_2$
Exact Mass: 556.16
Molecular Weight: 556.44
m/z: 556.16 (100.0%), 557.16 (37.3%), 558.16 (7.6%), 554.16 (6.4%), 555.16 (2.3%), 557.15 (1.5%)
Elemental Analysis: C, 69.07; H, 5.07; Fe, 10.04; N, 10.07; O, 5.75

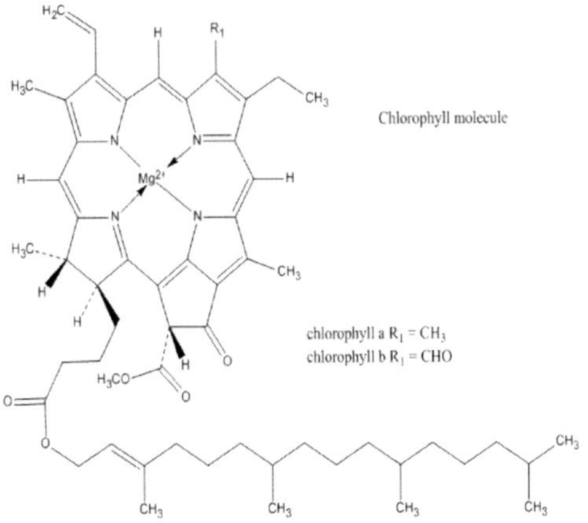

Chemical Formula: $C_{55}H_{71}MgN_4O_5^{2-}$
Exact Mass: 891.53
Molecular Weight: 892.48
m/z: 445.76 (100.0%), 446.26 (73.2%), 446.76 (41.1%), 447.26 (11.2%), 447.27 (4.2%), 447.76 (2.9%), 446.26 (1.5%), 447.77 (1.2%)
Elemental Analysis: C, 74.02; H, 8.02; Mg, 2.72; N, 6.28; O, 8.96

Regarding eye lens, the crystalline lens epithelial cells lose their nuclei as they become mature lens fibers, thereby, as red blood cells; both types of cell have neither nuclei nor mitochondria. The eye lens in humans is covered in the anterior part by the iris, a tissue heavily pigmented, which has 40 % more melanin than the skin (Fig. **6**).

There are some notable contradictions about the way that oxygen follows from blood to inner cell, for instance: in populations of mitochondria in cold-adapted animals are often so densely clustered that it is difficult to envision how oxygen might pass to those organelles deep within the cluster [3]. If mitochondria in the center of these clusters are not adequately supplied with oxygen, their functional efficiency will be impaired and therefore the functional significance is conflicting.

Unlike most normal tissues, a cancer cell tended to "ferment" glucose into lactate even in the presence of sufficient oxygen to support mitochondrial oxidative phosphorylation, and is termed the Warburg Effect. The inefficient use of sugar in the absence of oxygen was first noted by Pasteur in 1876. In the 1920s before ATP was discovered, Otto Warburg observed something unusual: Cancer cells and embryonic tissues accumulated lactic acid even in the presence of oxygen. It is difficult to understand that a cancer cell and embryonic tissues, whose energetic requirements become higher due to increased mitosis, it is going in for bypass oxidation of glucose in mitochondria, as if them are trying to prevent the complete oxidation of glucose and thereby the ATP production; anaerobic glycolysis produces only two ATP molecules, but somewhere between 30 and 36 ATP molecules are produced by oxidative phosphorylation; by other side, the demands for carbon incorporation biomass, that also requires energy; clearly supersedes the needs of ATP production from glucose. It is expected that the phosphate requirements are huge in cancer cell and embryonic growing tissues, due to, for instance, that cell membrane synthesis requires increasing amount of phosphate molecules for the buildup of phospholipids of the cell membrane and other organelles. Thereby, when the requirements of phosphate groups make smaller, then the role of ATP mechanism to keep the intracellular phosphate levels in the adequate range come into play.

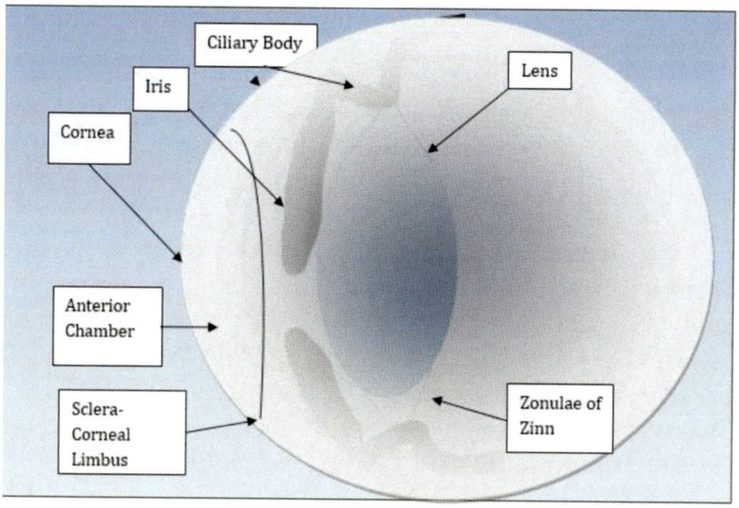

Fig. (6). The uvea is named for its high content of melanin, so its color is very similar to the grape. The transparency of the cornea, crystalline and vitreous depend on the energy that emanates from the melanin, because the transparency of the tissues represents a high energetic level stage.

The Not-possible Enzyme

Otto Warburg sought; as did other investigators, an enzyme that mat normally switch the gears from anaerobic to aerobic metabolism when oxygen becomes available. Such a simple enzyme, whose inactivation might be basic to the cancer process, has never been found. The switching mechanism is enmeshed in an intricate network of reactions that enables the carbon atoms and chains of the complete combustion of glucose to be largely conserved for other aims, as biomass replenishment and other vital functions as to keep phosphate groups level lowest by means of ATP synthesis.

Probably the reason for the apparent activity loss of several enzymes in the cancer cell is due to low levels of available chemical energy that coming from water dissociation and re-formation (Human Photosynthesis™) and not from glucose. We could say that cancer cell has a generalized failure, which is a characteristic of very low levels of free chemical energy, and could be schematized as follows:

Generalized failure of cell metabolism = Low levels of available chemical energy

Uterine and Umbilical O_2 Uptakes

The averages of the uterine and umbilical uptakes [4] had been shown that the mean fetal O_2 uptake was 45 percent % of the total. Consistently the utero-placental mass requires more O_2 than the fetus itself, in some cases 300 % more. The mean oxygen consumption "per kg of tissue" of the utero-placental mass was 14.1 ml S.T.P./min/kg, about twice the mean fetal O_2 consumption "per kg of tissue" which was 7.1 (in sheep fetus).

Supposing that glucose oxidation is the source of energy in mammals, it is illogical that the oxygen embryonic tissue consumption is less almost by half, than utero placental mass; given that embryonic tissues are growing at an accelerated rate, thereby almost complete glucose oxidation by mitochondria should be at a peak rate; and experimental measurements shown as blocked. By other side the pO_2 should be highest given that ox-phos requires significant amount of oxygen, however, pO_2 level in fetal blood that can be considered as arterial equivalent is 65 mm Hg at most. The values of pO_2 are paradoxical from glucose point of view, but in the light of Human Photosynthesis™, they are congruous: fetus and adult build the biomass arising from glucose, the perfect building block; therefore, the fetus take the glucose from mother blood and adult from meals; and more important, energy, defined as anything that produces a change, are taking from water in both cases, from ingested water in adults and from amniotic fluid in fetus, mainly.

In embryonic tissues the oxygen levels are lowest than mother´s tissues normally. Charles Darwin suggested that life may have begun in warm water with low levels of oxygen, and the amniotic fluid is characteristically warm water with low levels of oxygen. The history repeats itself and in everyone that´s how it all began.

Water and Glucose

The driving of glucose in human body requires important amount of energy and water. For instance: the kidney can filter and reabsorbs approximately 375 mg of glucose per minute, processes that undoubtedly requiring vast amounts of available chemical energy (from melanin)and water in order these processes can be accomplished adequately.

The Source of O_2 for the Cell

The O_2 required for oxidative phosphorylation is primarily delivered by the blood [5], however, from a mechanistic standpoint, this route must overcome important constrains.

Changing temperature affects both the solubility (α_{O2}) and diffusion coefficient (DO_2). DO_2 in water increases by approximately 3% for each rise of 1 °C. In light with Henry´s Law the solubility of oxygen decreases with increasing temperature (Sidell, 1998) by approximately 1.4 % $°C^{-1}$.

The reduction in DO_2 in the aqueous cytoplasm of cells with decreasing temperature is attributable to two factors: 1) a reduction in the kinetic energy of the system and 2) an increase in cytoplasmic viscosity, which is inversely related to cell temperature being the viscosity of aqueous cytoplasm the determinant factor of intracellular diffusion.

The effect of temperature on the viscosity of aqueous solutions is considerable. Between 5 and 25 °C, a thermal range that includes the physiological temperature range of many temperate-zone fish species, the dynamic viscosity of pure water increases by more than 70%, from 0.89 to 1.52 centipoises. Thereby, diffusion of dissolved oxygen in the aqueous compartment of the cell is quite sensitive to decreasing temperature primarily by the elevation in cytoplasmic viscosity.

The Cell Can Produce O_2 at its Own, *In Situ*

There are real possibilities that intracellular Oxygen (O_2) comes mainly from the dissociation and re-formed of the water molecule due to it is a process that happens constantly in every single cell of the body, this is: *in situ* (Fig. 7).

In accord with the fundamental photo-chemical reaction of the Human Photosynthesis™, the Solís-Herrera™ cycle can be schematized as follow:

$$\mathbf{2H_2O \leftrightarrow 2H_2 + O_2 + 4e^-}$$

The dissociation and re-formed of the water molecule takes place in the perinuclear space, thereby the diatomic hydrogen carrying the energy and the molecular oxygen (O_2), one of the most stable molecules, are released symmetrically, spreads out in all directions, as growing spheres with a center of melanin. One sphere is composed from dissociated water and the next one from re-formed water.

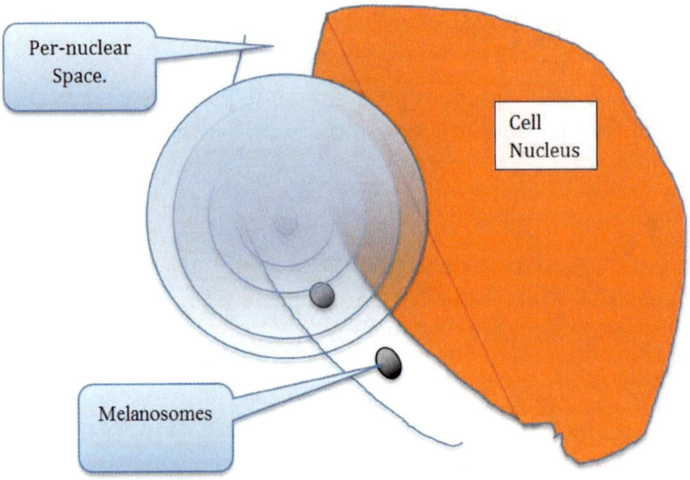

Fig. (7). The melanosomes are found mainly in the Perinuclear space, surrounding the cell nucleus, forming a kind of wrapper. This enveloping arrangement explains the source of energy of the cell nucleus, since under normal conditions it does not contain mitochondria or ATP.

Oxygen is Toxic at Any Level

Perhaps the oxygen biological role in human physiology must be rethought, otherwise, contradictory reports as: "*Although oxygen supplementation is necessary for survival in many preterm infants, several studies have shown that it increases the risk of retinopathy of prematurity, bronchopulmonary dysplasia, periventricular leukomalacia and cerebral palsy* [6]" cannot be explained adequately.

Furthermore, the studies of Kassissia *et al.* in regards the kinetics of tracer oxygen distribution, shows that the proportion of the tracer oxygen (^{18}O) returning to the circulation from tissue is a small fraction of the total tracer emerging at the outflow [7].

The observation about that melanin content is inversely related to mitochondria number until an 83 %; is congruous with, for instance: the fact that thermal acclimation of fishes can induce very dramatic changes in subcellular anatomy. During acclimation to cold temperature, from 28 to 2 °C, of crucian carp (Johnston and Maitland, 1980) the mitochondrial density of red muscle fibers increased by 80 %; a finding that was documented also in numerous other fish species. The percentage of cell volume displaced by mitochondria changes from 28.6% at 25 °C to 44.8 % at 5 °C (Egginton and Sidell, 1989).

In the light of Human Photosynthesis™ unraveling, we could say that when cold weather turn down the turnover rate of water dissociation and re-forming; then the mitochondria number increases as adaptive process but not as accessory source of energy, instead is to optimize other mitochondria functions as heat-generating mechanism in non-shivering thermogenesis in young mammals; and it has also been suggested that the production of free radical species by mitochondria may play a key role as signaling mechanism [8]; a regulator of phosphate groups and calcium. In our opinion mitochondria has important functions but in relationship with biomass, being a regulator mechanism that contributes to the fine-tuning balance between energy and biomass, developed by nature along 4 billion years of evolution, but mitochondria function depends of the energy that comes from melanin´s water dissociation (Fig. **8**).

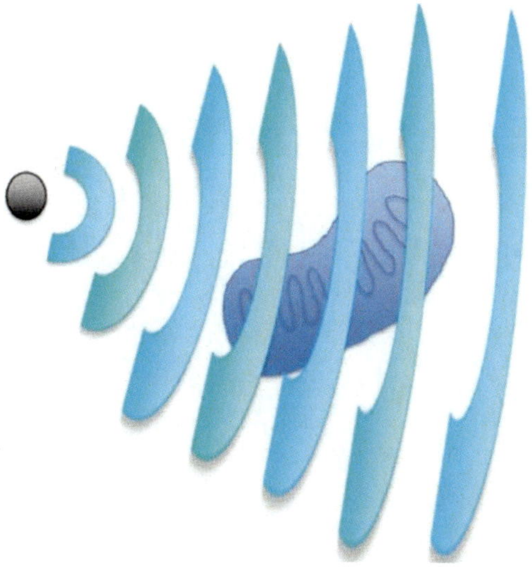

Fig. (8). Melanin (dark dot) is the real source of energy of mitochondria, due to its previously unknown intrinsic property to split and re-form the water molecule, releasing therefore free chemical energy. Energy is released symmetrically in all directions, as growing spheres of alternate nature, one sphere has a high content of diatomic hydrogen, which does not combine with water; and the next growing sphere has re-formed water and a flow of high energy electrons, that did not combine with nothing. The physiology of mitochondria must be rethought considering this completely different source of energy.

It is expected some already known relationship of light with mitochondria, for instance; the illumination-induced mitochondrial depolarization. It is not surprising that the transient openings of mPTP (Mitochondrial Permeability Transition Pore) appear to be completely innocuous and cause no evident harm to the cell, with no sign of apoptosis or cytochrome C release in illuminated cells even 24 h later [9]. In cardiomyocyte and cortical astrocyte, the first response to illumination consists of rapid short-lived transient and localized depolarization of mitochondria.

The mitochondria proton-motive-force is expressed as membrane potential and is usually estimated as some 150-180 mV negative to the cytosol, which is the stored force that underpins the production of ATP. Once has been synthesized, ATP must then be transported into cytosol by the adenine nucleotide translocase (ANT); and this is an argument opposed to ATP as universal currency of energy for the cell, because the terminal phosphate is added and removed three times each minute [10]; instead the regulation of biomass, for instance: phosphate groups regulation are more plausible, due to the mitochondrial potential, usually referred to as $\Delta\psi_m$, also drives ATP synthesis and too the accumulation of calcium into mitochondria. Calcium that is accumulated by mitochondria must be removed, and this is achieved largely through the action of a Na/Ca exchanger that undoubtedly requires energy and probably from melanin.

2. - Performing various chemical reactions that use nutrients and O_2 to provide energy for the cells, traditionally (and wrongly) as follows:

$$\textbf{Food} + O_2 \rightarrow CO_2 + H_2O + \textbf{energy}$$

It is a long-lasting misleading that the source of energy for the body is the chemical energy stored in the carbon bonds of ingested food. Dietary carbohydrates are broken down primarily into glucose; thereby the anterior reaction can be schematized as follows:

$$\textbf{Glucose} + O_2 \rightarrow CO_2 + H_2O + \textbf{energy}$$

If glucose were a source of energy, then diabetic patients should be able to fly. Therefore, the reaction must be rethought as follows:

$$\textbf{Glucose} + H_2 + O_2 \rightarrow CO_2 + H_2O + \textbf{biomass}$$

Note: H_2 and O_2 are provided by the melanin´s water dissociation.

The Uniqueness of Melanin

Melanins are biologically very important and little understood molecules, their physicochemical properties are frustrating and much remains to be learned (Magno, Joris; 2001). These were the general perception about melanin until before our finding of the intrinsic property of melanin to split and re-form the water molecule. Melanin works well in pure water, a key difference with enzymes, which free in solution did not work.

There are several varieties of melanin; all are polymers of hydroxy-aromatics. They are very ancient molecules, from the beginning of the times; and extremely stable: melanin had been found in the ink sacs of fossilized squid that died 180 million years ago [11]. Eumelanin is present not only in the skin and hair but also in the retinal pigmented epithelium, choroid layer; choroid plexuses, (choroid means grape color); also in the meninges, in the substantia nigra, locus coeruleus, in the stria vascular is of the inner ear; in the chromaffin system (adrenal medulla, sympathetic ganglia), and perhaps in every single cell of the body, where the intracellular amount depending on location and function of the cell itself, for instance: more energy requirements more melanin.

The pigment in the skin is produced by the melanocytes and packed in granules termed melanosomes in mammals and melanophores in amphibians; that are membrane-bound and have a characteristic structure. These melanosomes are transmitted to the surrounding epithelial cells (keratinocytes) in a manner that may be unique in biology: each melanocyte actively implants its granules into the body of the cells close to it, requiring thereby energy. Under pathologic conditions this activity can be disturbed resulting in poor pigmentation.

Characteristically the melanosomes surround the nucleus, presumably to shield its DNA from radiation; however, at light of the intrinsic property of melanin to split and re-form the water molecule; the function seems different and obvious: melanin is the source of energy of the cell nucleus (Figs. **9** and **10**).

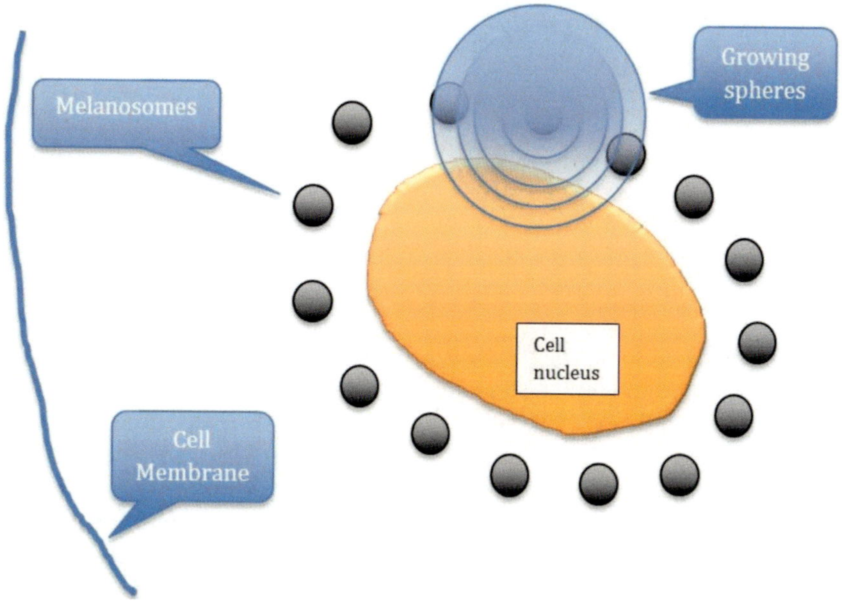

Fig. (9). The melanosomes characteristically are located surrounding the nucleus. The growing spheres of energy released usually overlapping and all them are coalescing in the middle area of nucleus, forming, thereby a zone of high energy, that is crucial for the nucleus functioning.

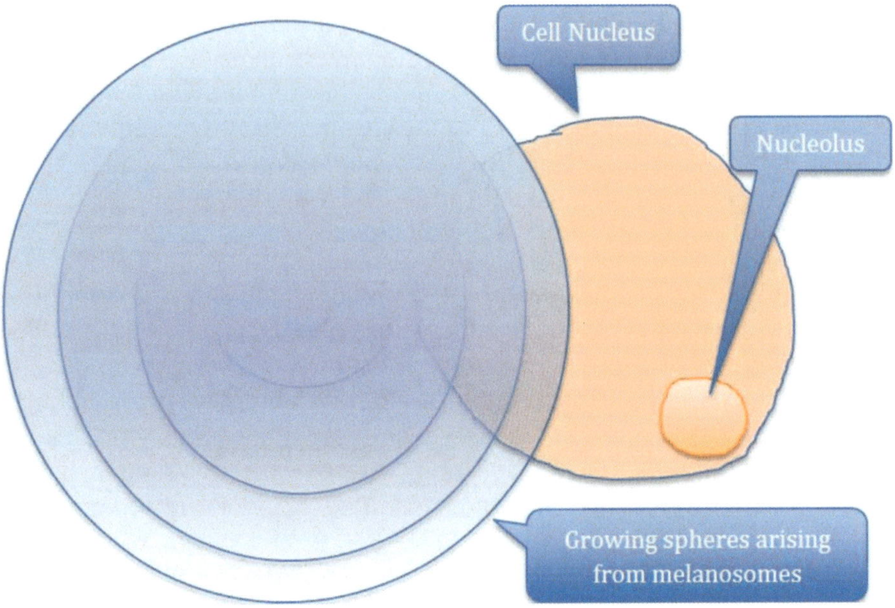

Fig. (10). The cell nucleus has not mitochondria, and ATP molecules are not able to reach the inner side of the largest intracellular organelle. Energy requirements of the nucleus are indeed huge. The question is: Energy for the nucleus, from where? The answer: from melanin of course.

Chemically, eumelanin is an insoluble polymer of tyrosine, and pheomelanin, the typical pigment of the red hair; is an alkali-soluble polymer of tyrosine and cysteine. The molecular weight of melanin is unknown so far.

The redox properties of melanin have long been used in histochemistry. At pH 4, melanin granules reduce silver nitrate to metallic silver, constituting the classic method for the histochemical detection of melanin and are called the Fontana Reaction. Plants melanins are catechol melanins, unlike mammalian melanins, which are indol melanins [12].

The specific ultrastructure of melanin, in spite great technological advances; cannot be determined. In the Central Nervous System, melanins is not segregated in specific pigment cells but resides in neurons where it is contained in specific bodies, wrongly considered residual bodies, instead are melanin bodies are highly complex structures with fundamental functions, the main one: energy production. It is present in a row of nuclei along the brainstem; the total amount of neuromelanin tends to decrease with age, apparently due to the neuromelanin-containing neurons decrease in number; however, in populations with skin photo-type V, the amount of melanin in substantia nigra is even less than in a similar cohort of Caucasian population with skin photo-type I, II or III.

Melanins and catecholamines are booth derivatives of tyrosine (Fig. **11**). Interestingly albinos do have neuromelanin, although they have little in other parts of the body, among them in the skin and eye, usually they have eyesight problems. By other side, carnivores, including man, have the largest amount of neuromelanin and rodents little or none apparently, however the explanation could be that in small species, the size of the midbrain and therefore of the mammal, is a determining factor in the mesencephalon pigmentation, in other words: larger animals require more energy in midbrain, smaller animals require less energy for an adequate midbrain functions [13].

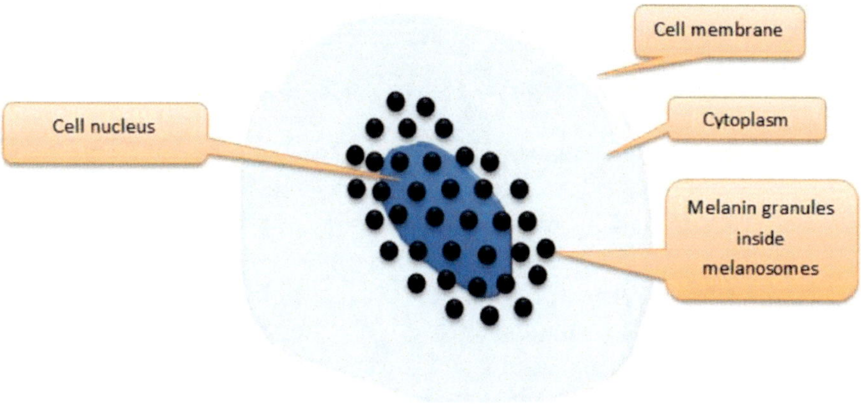

Fig. (11). Melanosomes normally surround the cell nucleus forming a wrapping-like structure, that cover the largest intra-cell organelle at full. This arrangement ensures that cell nucleus is permanently immersing in these energy growing-spheres that emanate from melanin. The energy for the nucleus is a case because the growing spheres coalescing in the central nucleus region, forming, thereby; a high energy zone, constituting the source of energy by excellence for the cellular nucleus. Recall that cell nucleus neither has mitochondria nor ATP.

Melanin and lipofuscin can be associated. About 70 % of the brown pigment in the myocardium is lipofuscin and 30 % is melanin. These pigments are associated also in the nerve cells, the liver and possibly other organs.

The already much-described, but not easy to proof functions of eumelanin are:

a. Light absorption: Eumelanin absorbs throughout the ultraviolet and visible regions of the spectrum, but in accordance with our research melanin has the amazing capacity to absorb the entire electromagnetic spectrum, from radio-waves to gamma rays.
b. Melanin converts light to heat. This is an ancient mistake based in that practically any already known substance is able to absorbs light and then dissipates it in form of heat, however melanin is unique and therefore has a unique behavior in that sense. Melanin is the only known substance able to

dissipate the absorbed energy in a non-irradiative way, and this is by mean of the dissociation of the water molecule, an expensive process from the energetic point of view, recall that in the laboratory requires 2000 °C.

c. Melanin is a controversial radical sink. It is based in the free-radical theory of aging and in the observation that black skin aged better than white. The controversy arising from the fact that eumelanin produces potentially damaging free radicals when exposed to light [14]. Our point of view is that melanin produces the best antioxidant: The Molecular Hydrogen; and this is a fact out of discussion. An indirect proof is that most of our tissues maintain their intracellular oxygen concentration in the micro-molar range, 1000 times below the bloodstream value [15].
d. The reaction inside melanin can be schematized as follows:

$$\mathbf{2H_2O \leftrightarrow 2H_2 + O_2 + 4e^-}$$

Thereby, the main product is diatomic hydrogen, the energy carrier by excellence in the entire Universe; by other side diatomic oxygen is a very stable molecule. This reaction is performed by melanin under normal conditions, these are when melanin is healthy, and the amount of water and visible and invisible light is sufficient. Oppositely, when melanin is poisoned by heavy metals as iron, cadmium; mercury; *etc.*, then the basic reaction is impaired, and the production of free radicals will happen, because instead of diatomic hydrogen and oxygen, then we come to have monoatomic hydrogen and oxygen besides hydroxyl molecules. Recall that in presence of metals or Boron, hydrogen works with -1.

a. Ion-exchange resin: Melanin granules do behave as lumps of ion-exchange material but whether this amount to a useful function depends of the circumstances, substances that get involved and the surrounding. For instance, when melanin works as an electron acceptor, bind many drugs, which act as electron donors. Among these are cocaine, epinephrine, phenothiazine –an antidepressant- which blocks dopamine receptors, and specially the antidepressant chlorpromazine, that changes substantially the electrical behavior of melanin and is used in the treatment of amphetamine poisoning.

The small amount of mitochondrial DNA codes for only 13 polypeptides in humans. These are mainly the hydrophobic cores of the major trans-membrane proton pumps, which are sticky and insoluble and difficult to move around the cell. All the remaining mitochondrial genes have migrated to the nucleus, and the hundreds of other mitochondrial proteins are now imported from the cytosol. Surprisingly, despite mitochondrial DNA is less well protected than nuclear DNA; significant mitochondrial variants are not observed in practice. In healthy people all the copies of the mitochondrial DNA are identical.

Mitochondria are subcellular organelles containing the Krebs cycle, fat oxidation pathway and the respiratory chain, which produces almost all of the 70kg of ATP, used each day in the human body. They can be purified from tissue homogenates by differential centrifugation, using non-penetrant isotonic media to provide osmotic support. A few tissues (*e.g.* red blood cells, eye lens) have no mitochondria, and cannot respire aerobically. Summarizing, the source of energy of the mitochondria is the Human Photosynthesis™.

The finding of the intrinsic property of melanin to split and reform the water molecule is a conceptual revolution of biblical proportions [16]. The entire physiology of the cell must be rethought. Fortunately, Human Photosynthesis™ will not change the Earth rotation movement, or the moon phases; only is modifying our conception about life.

CONCLUSION

Melanin was the missing link in the sequence of the origin of life. Now that we understand that the energy required by the Eukaryote cell comes from melanin, we can understand the physiology of the body in an unusual way.

For example, the natural history of melanin is very consistent with aging. The ability to transform light into chemical energy by dissociating the water molecule begins to get lost at 26 years, by about 10% every ten years, and after the 50s goes into freefall.

This is congruent with the observed fact that from that age (26 years), we began to lose bone mass, muscle mass, our skin is thinning, *etc.* It is an axiom that any system that loses energy loses mass.

The physiology of our body is completely governed by the generation and distribution of energy that comes from melanin. They are surprisingly accurate processes that are altered with contaminated water, polluted air, pesticides, herbicides, alcohol, drugs, addictions, *etc.*

Even the affective problems have to do with the unchanging order regarding the generation and distribution of energy. The proof of this is the seasonal depression that occurs in the low-light months.

When the body goes into balance, so does the mind. Any cell or tissue in our body requires light, and the amount is strictly regulated, hence in the cold countries, the amount of melanin in the skin is lower so that it can penetrate the right amount of visible and invisible light so that even the bone, which also needs light, to cover your energy needs.

The discovery of the true source of energy of the human body opens a new era in biology; and aging is no exception. Health problems related to advancement in age, such as Alzheimer's, Parkinson, osteoporosis, cancer, degenerative osteoarthritis, sarcopenia, frailty, *etc.*, now have new real options for prevention and treatment: the balance between mass and energy.

HUMAN AND ANIMAL RIGHTS

No humans were used in this research. All animal research procedures followed were in accordance with the standards set forth in the eighth edition of Guide for the Care and Use of Laboratory Animals published by the National Academy of Sciences, The National Academies Press, Washington, D.C.

CONFLICT OF INTEREST

The author (editor) declares no conflict of interest, financial or otherwise.

ACKNOWLEDGEMENT

This work was funded by Human Photosynthesis® Research Center.

REFERENCES

[1] Solis-Herrera, A. The origin of life according to Melanin. *MOJ Cell Sci. Rep.*, **2018**, *5*(1), 00105. [http://dx.doi.org/10.15406/mojcsr:2018.0500105]

[2] Zhanf, M.; Li, W.; Niu, G.; Leak, R.K.; Chen, J.; Zhang, F. ATP induces mild hypothermia in rats but has a strikingly detrimental impact on focal cerebral ischemia. *J. Cere.Bl Flow Met.*, **2013**, e1-e10.

[3] Sidel Bruce, D. Intracellular oxygen diffusion: The roles of myoglobin and lipid at cold body temperature. *J. Exp. Biol.*, **1998**, *201*, 1118-1127.

[4] Giacomo, M.; Cotter, J.R.; Makowsky, E.L.; Barron, D.H. Simultaneous measurement of uterine and umbilical blood flows and oxygen uptakes. *Q. J. Exp. Physiol.*, **1966**, *52*, 1-18.

[5] Lauralee, S. Human Physiology, from Cell to Systems. In the Chapter Muscle Physiology. Thomson Brooks/Cole. 5th ed. Autralia, Canada, Mexico. **1990**, Fifth Edition. p. 277.

[6] Vaucher, Y.E.; Peralta-Carcelen, M.; Finer, N.N.; Carlo, W.A.; Gantz, M.G.; Walsh, M.C.; Laptook, A.R.; Yoder, B.A.; Faix, R.G.; Das, A.; Schibler, K.; Rich, W.; Newman, N.S.; Vohr, B.R.; Yolton, K.; Heyne, R.J.; Wilson-Costello, D.E.; Evans, P.W.; Goldstein, R.F.; Acarregui, M.J.; Adams-Chapman, I.; Pappas, A.; Hintz, S.R.; Poindexter, B.; Dusick, A.M.; McGowan, E.C.; Ehrenkranz, R.A.; Bodnar, A.; Bauer, C.R.; Fuller, J.; O'Shea, T.M.; Myers, G.J.; Higgins, R.D. SUPPORT Study Group of the Eunice Kennedy Shriver NICHD Neonatal Research Network. Neurodevelopmental outcomes in the early CPAP and pulse oximetry trial. *N. Engl. J. Med.*, **2012**, *367*(26), 2495-2504. [http://dx.doi.org/10.1056/NEJMoa1208506] [PMID: 23268664]

[7] Kassissia, I.G.; Goresky, C.A.; Rose, C.P.; Schwab, A.J.; Simard, A.; Huet, P.M.; Bach, G.G. Tracer oxygen distribution is barrier-limited in the cerebral microcirculation. *Circ. Res.*, **1995**, *77*(6), 1201-1211. [http://dx.doi.org/10.1161/01.RES.77.6.1201] [PMID: 7586233]

[8] Duchen, M.R. Roles of mitochondria in health and disease. In Section III: Mitochondria, β-cell function, and type 2 diabetes. *Diabetes*, **2004**, *53*(Suppl. 1)

[http://dx.doi.org/10.2337/diabetes.53.2007.S96]

[9] Jacobson, J.; Duchen, M.R. Mitochondrial oxidative stress and cell death in astrocytes-requirement for stored Ca^{2+} and sustained opening of the permeability transition pore. *J. Cell Sci.,* **2002**, *115*(Pt 6), 1175-1188.
[PMID: 11884517]

[10] Arthur, K. *For the Love of Enzymes*; First Harvard University Press: Cambridge MA, **1991**, p. 65.

[11] Guido, M.; Isabelle, J. Chapter 3Symptoms of Cellular Disease: Intracellular Accumulations.*Cells, Tissues, and Disease. Principles of General Pathology,* First Edition; ISBN 0-86542-372-5.

[12] Jimbow, K.; Fitzpatrick, T.B.; Wick, M.M. Biochemistry and physiology of melanin pigmentation.*Biochemistry and Physiology of the Skin*; Goldsmith, L.A., Ed.; Oxford University Press: New York, **1983**, pp. 687-712.

[13] Aliev, G.; Solis, H.A.; Li, Y.; Kaminsky, Y.G.; Yakhno, N.N.; Nikolenko, V.N.; Zamyatnin, A.A. Benberin, Valery. Bacurin, Sergey O. Human Photosynthesis™, the Ultimate Answer to the long-term Mystery of Kleiber's law or $E = M^{3/4}$: Implication in the context of gerontology and neurodegenerative diseases. *O. J. Psych.,* **2013**, *3*, 408-421.
[http://dx.doi.org/10.4236/ojpsych.2013.3405]

[14] Sealy, R.C.; Felix, C.C.; Hyde, J.S.; Swartz, H.M. *Structure and Reactivity of Melanins: Influence of Free Radicals and Metal Ions. Free Radicals in Biology*; Academic Press: New York, **1980**, *4*, pp. 209-259.

[15] Bioenergetics. Origin & evolution of mitochondria. Biochemistry & Molecular Biology at the University of Leeds: http://www.bmb.leeds.ac.uk/illingworth/oxphos/evolve.htm

[16] Arturo, S-H. The Human Photosynthesis™. Infinity Publishing Co. **2013**.

CHAPTER 9

The Unexpected Role of Melanin as Bio-energetic Molecule and Prof. Szent-gyorgyi´s Energy Concepts: Hydrogen as Main Source of Energy of Muscle Cell

Sergey Suchkov[1], Arturo Solís-Herrera[2,*] and Ruth Isabel Solís-Arias[3]

[1] *I.M. Sechenov First Moscow State Medical University (FMSMU), Moscow, Russia. A.I. Konstantinov Moscow State Medical & Dental University, Moscow; Russia*

[2] *Director and Founder, Human Photosynthesis® Research Center, Aguascalientes, México*

[3] *Member Clinical Staff, Human Photosynthesis® Research Center, Aguascalientes, México*

Abstract: The reason for writing this article is to deepen in two concepts that we believe are fundamental in the build-up of a new Biochemistry and therefore a new era in medicine, because allow us a better understanding of the biology of the eukaryotic cell, and specifically for this chapter, of the muscle cell.

The first one is the theoretical concept of intracytoplasmic cell´s common energy levels proposed by Prof. Szent-Gyorgyi in the first half of the past century and the second one the surprising fact that the muscle relaxation involves higher levels of chemical energy than the muscle contraction stage, which presupposes a reduction in the level of the energy. The hitherto intrinsic chemistry of melanin to dissociate the water molecule fits and finally explains perfectly the teachings of Prof. Szent-Gyorgyi, and allows, at last, to move them to clinical practice.

The physiology of contraction and muscle relaxation has tried to explain based on glucose and ATP as main sources of muscle power, but has not been achieved satisfactorily despite notable efforts of researchers. And proof of this is that muscle problems, even the most common, will have not been able to improve in clinical practice.

The finding about the unsuspected function of melanin to generate chemical energy dissociating the water molecule, opens a new and promising panorama in the muscle cell biology. Which is complemented by our recent finding that molecules which

* **Corresponding author Arturo Solís-Herrera:** Human Photosynthesis® Study Center, Av. Aguascalientes Norte 607, Pulgas Pandas Sur, Aguascalientes, México; C.P. 20138; Tel/Fax: 524499160048/524499150042; E-mail: comagua2000@yahoo.com

Arturo Solís Herrera (Ed.)

contain the Heme group, such as chlorophyll, Myoglobin, hemoglobin, CYT P450, and the bilirubin, are also able to dissociate the water molecule but, unlikely melanin; irreversibly.

Keywords: Aging, Energy, Hydrogen, Melanin, Muscle disorders, Oxygen, Water dissociation.

INTRODUCTION

Prof. Albert von Szent-Gyorgyi (b.1893 Budapest, d.1986), during his studies about chemistry of cell respiration, described the interdependence of oxygen and hydrogen activation. He also demonstrated the existence of a reducing substance, later identified as vitamin C; in plant and animal tissues. In 1938 he commenced work on muscle research and discovered the proteins actin and myosin and their complex. This led to an empirical reproduction of muscle contraction which formed the trend of muscle research in the following decades.

His remarks were of a unique precision, and probably it was unable to reach deeper because at the time melanin was still considered as a simple sunscreen. The old concept that eukaryote cell could not transform light energy into chemical energy, is reflected in the following sentences taken the article entitled Bioelectronics, written by Prof. Szent-Gyorgyi and published by the journal Science in September 1968, about interactions between biological substances and light, observed *in vitro*, however, he thinks that charge transfer seems as have not a role in the biochemistry of biology.

While detected significant changes in organic matter in the presence or absence of light, Prof. Szent-Gyorgyi chose to make them aside, even trying to explain them, he discarded flat them writing that the idea of transfer of charges could not have a role in biology. He thought that strong charge transfer could play no role in biology, because the presence of strong oxidizing agents is incompatible with life.

And the professor had all the reason, but the dissociation of water as consistent and surprisingly accurate that occurs inside of the melanin, where liquid water turns into its components gaseous, producing H_2 and O_2, and by the way, the best antioxidant called is molecular hydrogen.

Finally, Prof. Szent-Gyorgyi believed that we have no light in our body to move electrons (except in the eye and skin), however visible and invisible light can reach all tissues [1]. So “spontaneous” charge transfer seems, for the biologist, a chemical curiosity, except if it is a highly precise process, as happens in melanin.

Therefore, the expression: “we have no light in our body to move electrons”, is not accurate due to visible and invisible light is already known that can penetrate all mammalian tissues to a considerable depth; for instance, it has even been detected within the brain of a living sheep [Wurtman, 1975], furthermore, melanin is the molecule that absorbs those electromagnetic radiations dissipating them through the splitting of the water molecule.

Using relatively advancing methods as electron spin resonance, Prof. Szent-Gyorgyi shown that even molecules apparently stable, whose key role is as intermediate metabolites or hormones, are, under determined circumstances, able to give off a whole electron, forming a free radical; which suggest that an unexpected orderly charge transfer impelled by light, could be more common and fundamental biochemical mechanisms in cells and body than previously thought.

However, he wrongly interpreted them because of the biochemical processes that constantly occur in the body, and not precisely as result of interaction between light and organic matter. But the free radicals that are formed during normal biochemical processes or even by the interaction direct between the electromagnetic radiation and the matter, are too random to give origin or at least to preserve the life. It is necessary the presence of a molecule capable to impose the order, and in case of mammals, is melanin, that is a strong electron acceptor [2].

It is not by chance that, in some molecular complex and during biological reactions, an electron can pass of a molecule to another, the first acting as acceptor and the second as receiver, what we conceptualize as redox.

But are processes highly ordered, surprisingly accurate, very homogeneous, that have been repeated incessantly over millions of years. And such consistency so extraordinary, comes from the very first spark of life, that came from melanin, this is: the energy that started life and until today still constitutes the base of life.

Prof. Szent-Gyorgyi, believed that the components of the cell can be important sources of electrons in possibility to be transferable, happen in elements as nitrogen, sulfur, and oxygen atoms, due to these elements have pairs of isolated electrons, this is: electrons which do not take part in bonding processes, and are, therefore, available for transfer under certain circumstances. However, the acceptors story is so different; due to cells are poor in these kinds of biochemical structures. Prof. Szent-Gyorgyi could find just one acceptor group, CO -carbonyl-, a kind of ketoid acceptor, which can accept in its double bond (C=O) an additional electron. However, as acceptor molecule, CO is too weak. Opposite, melanin is a strong electron acceptor, which never was considered by the Prof Szent-Gyorgyi; instead he mentioned that two CO groups, *i.e.*, in an aromatic molecule, at least theoretically, must make a very good acceptor.

A living cell requires a constant flux of chemical energy aimed not only for all its diverse and complex functions, even also for the maintenance and/or replenishment of its morphological structure. Without an adequate source of chemical energy life would be extinguished almost instantaneously, and the highly complex cellular organization would collapse. The main source of this energy is undoubtedly the sunshine.

Recent investigations, especially during the last years have brought hydrogen in the place of carbon, and instead of CO_2 seems water as the origin, the constituent and fundamental support of any kind of life, into the foreground. For the Prof. Szent-Gyorgyi, the foodstuff, carbohydrates, lipids, and proteins; were essentially as a hydrogen´s packets, likely a hydrogen supplier or a hydrogen donor, this is: hydrogen as energy carrier; and the main event during combination of oxygen with glucose (combustion, oxidation) theoretically is the separation of hydrogen from glucose molecule.

But the supposedly graduated combustion is a messy process that not only releases hydrogen monoatomic, but also radical oxygen, which are toxic. It is hard to imagine that life arose from the generalized disorder, so either because of a “burn” ranked and deemed to oxidative phosphorylation.

This progressive oxidation of glucose, in highly tuned stages seems to be the basic principle of oxidation in biology [3]. Thereby, water is not only the great solvent, also is: the matter and matrix, mother and medium of any kind of life, and unexpectedly is also source of energy. Simply there is no life without Light, Melanin, and Water, in order of abundance in the universe.

In regards the photosynthetic unit in plants proposed by Emerson and Arnold, Prof. Szent-Gyorgyi said that a photon is absorbed someplace in a group of already 300 light absorbing-molecules with chlorophyll acting as central molecule; and this energy is then transmitted in some unknown way, to an enzyme system where it is optimized. But evolutionarily it makes little sense because it would be difficult to explain how it is that the first 300 molecules of chlorophyll were formed before to capture the first photon, and about the highly complex subsequent enzyme system nor to say.

Despite the apparent simplicity of the measurements of chlorophyll fluorescence, the underlying theory and the interpretation of data remains complex and, sometimes, controversial [4]. The principle underlying chlorophyll fluorescence analysis is relatively straightforward. Light energy absorbed by chlorophyll molecules in a leaf can undergo one of three fates: 1) it can be used to drive photosynthesis (photochemistry), 2) excess energy can be dissipated as heat or 3) it can be re-emitted as light-chlorophyll fluorescence, being a small part of total light absorbed (1 or 2%). These three processes occur in competition, such that any increase in the efficiency of one will result in a decrease in the yield of the other two. Hence, by measuring the yield of chlorophyll fluorescence, information about changes in the efficiency of photochemistry and heat dissipation can be gained.

The spectrum of fluorescence is significantly different to that of absorbed light, with the peak of fluorescence emission being of longer wavelength than that of absorption. However, there is confusion factor as an unavoidable light loss that complicates measures.

Hydrogen as the Real Source of Energy of the Muscle

We will then describe some very interesting findings of the Prof. Szent-Gyorgyi which, to date; could not be explained adequately, and thereby cannot be translated to clinical practice. We believe that our discovery of the unexpected intrinsic ability of melanin molecule to transform light, visible and invisible, into chemical energy through of the breaking of the water molecule [5], it will mark a before and an after in the physio-pathology of diverse diseases that affect the muscles; which will allow structuring a preventive, predictive and personalized medicine more efficient than it has been so far [6]. Prof. Szent Györgyi works in a biochemical context of the first half of XX century. Being the main challenge in that time, the understanding of the nature of the energy coupling between organic matter and energy. The prevalent question in that time, was how is it possible that living organisms could capture available energy, coming from the degradation of organic matter, or better, from the absorption of light, and further, which way organisms harness it at every moment keeping the performance of useful work such as complex and accurate biosynthesis, also membrane transport, and finally movement. In principle, it is not feasible to explain how the cell takes up the energy of food and transforms them into work, since the chemical energy of the cells apparently does not comes from the light energy, due to, between other things, there were not known a transducer. Fritz Lipmann proposed that the source of immediate energy for the cell was the ATP or any phosphoryl donor involved chemically in the reaction in which participates. Theoretically, generation and maintenance of ATP supply, are functions of the main metabolic highways (*i.e.* respiration and photosynthesis). However, under our experimental conditions, ATP is not for purposes of energy, instead for control of the levels of phosphate; that are compounds characterized by thermodynamic instability but with a remarkable kinetic stability [7].

Being Chemiosmotic theory fans, mitochondria, supposedly; generate the major part of ATP through oxidative phosphorylation. Conceptualized as a metabolic pathway molded by a structure of redox proteins, often associated with membranes; that can funnel electrons from NADH, reduced in cytoplasm; to oxygen through a relatively impermeable mitochondrial membrane, and whose explanation is still a theory, due to mitochondria membrane is impermeable to the passage of the NADH produced by the extra-mitochondrial glycolysis. Theoretically, the chemical energy available from this exergonic and apparently incessant reaction is sufficiently conserved along several theoretical processes, and finally appears to be enough to drive the endergonic synthesis of ATP arising from ADP and inorganic phosphate, at least theoretically.

However, it is biochemically more congruous that ADP to ATP step is more exergonic than endergonic, releasing chemical energy; and not when ATP goes to ADP. It is also interesting that ATP synthase, in its usual membrane-bound form it can generate ATP, whereas the solubilized enzyme (free in cytoplasm) can only break it down.

This could mean that membrane-bound enzyme uses mainly high energy electrons and solubilized enzyme uses mainly diatomic hydrogen as source of energy respectively. Thereby, the question that remains about how mitochondria could transduce the chemical energy that respiration releases; making it suitable to drive ATP synthesis; being the most expensive the thermodynamic hill or energy of activation. Related questions had remained with respect to uptake of diverse metabolites by organelles and cells, about photo-phosphorylation.

And these long last mysteries will continue whilst the dogma of ATP and thereby glucose as energy source still prevails, in an analogous way to phlogiston of XVI and XVII centuries. It is not possible to explain adequately something that is not congruous with physic-chemical laws and principles. Proof of this is the fact that in spite the best efforts that have been made to try to identify the hypothetical intermediate metabolite that transfer the energy of respiration to the ATP synthase has been unsuccessful for decades. Here were concepts which prevailed at the time Dr. Szent-Gyorgyi conducted his experiments on muscle that then we will see in detail. Thereafter, in 1950s, Peter Mitchell´s theory and whom never works in mitochondria, propose that redox complex chains and the enzyme ATP synthase both translocate protons and that are linked only through the proton motive force or proton current.

However, more than 60 years later, Mitchell´s chemo-osmotic theory is stand still a theory, and not a high-level bio-energetic concept, due to the date has not been possible to prove it satisfactorily.

The Structure and Chemistry of Muscle

The studies under light-microscope reveals that the muscle fibers are formed by the fusion of many individual muscle cells during development; and consists of small fibers, 30-50 nm wide in average and up several centimeters in length. Sarcolemma surrounds the fibers, being a thin sheath, containing in their interior the contractile matter. The periodic variations of the displaceable or contractile matter, that is homogeneous cross-wise but along the fiber axis, lengthwise, make the fiber resemble a roll of coins.

By other hand, the contractile matter of the muscle is an arrangement of complex protein, made up of several poly-peptide molecules; actin, myosin, titin, and nebulin. All them are present as very thin filaments of molecular dimensions. The molecular arrangement formed by these proteins is a rather soft, elastic gel which, especially at rest, when it is highly vulnerable, is easily damaged and disorganized mechanically.

A relaxed muscle reflects a stage of higher energy than the contracted muscle, probably due to it is a highly complex stage, being the relaxation a process surprisingly accurate, therefore, more complex any process, more energy is required, and not any kind energy, instead must be also precise in time and form.

By themselves, neither myosin nor actin can contract [8]. If brought together in a suitable ionic milieu, then they unite to form a complex: "actomyosin." Supposedly, per the concentration and the nature of ions present, this actomyosin may be charged (theoretically) by the ATP and dissociate reversibly into its two components: actin and myosin, or else it may be discharged and dehydrated excessively.

Due to the elongated way actomyosin particles are oriented, the shrinking will be not uniform, and the complete gel also will show a shrink in the direction of the molecules′ axis and expands at right angles to this direction. Actomyosin threads, representing a significant part of muscle mass, contracting under these conditions, become shorter and wider without changing their volume. During muscle contraction, each sarcomere shortens, bringing the Z discs closer together. There is no change in the width of the A band, but both the I band, and the H zone almost completely disappear. These changes are explained by the actin and myosin filaments sliding past one another, when the reaction takes place in the sarcomere, where the elongated actomyosin filaments form a continuous system, the shortening will be able to do work as lifting weights, or even develop tension under isometric conditions, a stage that is usually called "contraction" (Fig. **1**).

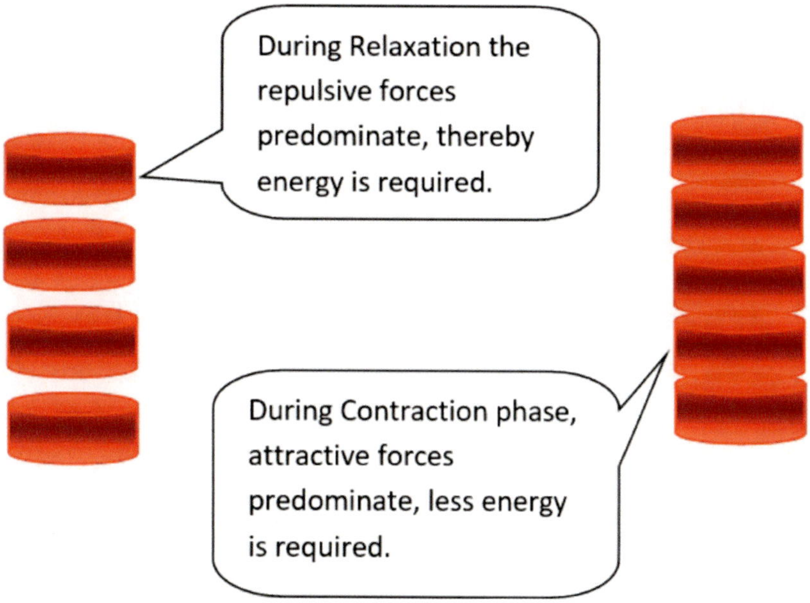

Fig. (1). Actin, myosin, titin, and nebulin; can schematized as piles of coins that move with relative ease depending on the available power levels. When the energy level is high, say basal; the repulsive forces of the protein complexes predominate, and coins "stacks" are separated in a surprisingly exact manner. When the levels of energy decrease, for example, when the ATP is transforms in ADP, the energy that absorbs conditions that the levels of energy chemical decrease, and then the forces of attraction predominate and then the "batteries of coins" is approach between itself.

Contractile matter is built of twisted functional units (*i.e.* actin/titin), and contraction is an "all-or-none equilibrium reaction" of these functional units, and are significantly dependent on temperature, pressure, reactants concentration, balance of charges; ATP presence, ions milieu; *etc.*

The importance of attractive and repulsive forces during muscular contraction and relaxation, is demonstrate by the extreme care in arrangement of polarization that Nature has in it (Fig. **2**). The thick filaments of muscle consist of several hundred myosin molecules, associated in a parallel staggered array by interactions between their tails. The globular heads of myosin bind actin, forming cross-bridges between the thick and thin filaments. It is important to note that the orientation of myosin molecules in the thick filaments reverses at the M line of the sarcomere. The polarity of actin filaments (which are attached to Z discs at their plus ends) similarly reverses at the M line, so the relative orientation of myosin and actin filaments is the same on both halves of the sarcomere.

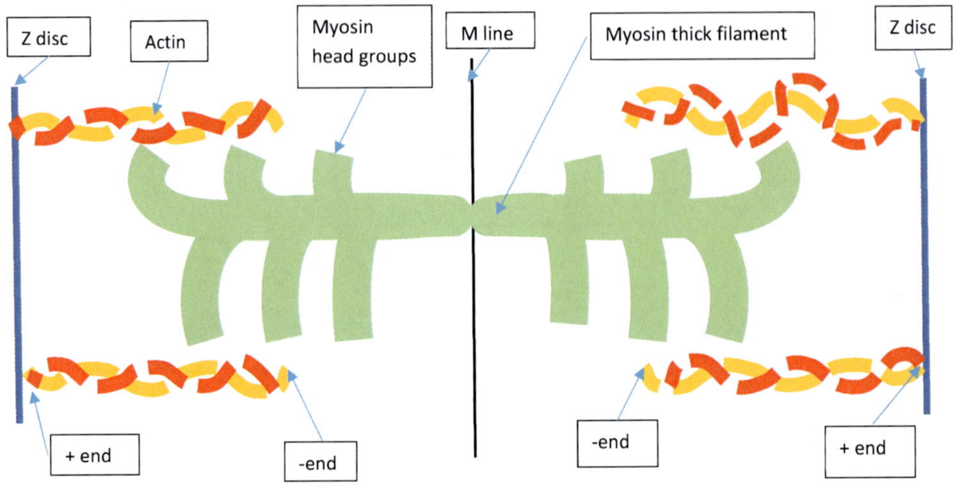

Fig. (2). Per our results, the muscle contraction is a stage of less energy than the relaxation. When the levels of chemical energy are diminished, attractive electro-magnetic forces dominate, by which the actin and myosin are attracted, as chemical energy rises, then the repulsive electro-magnetic forces dominate, which induces separation of actin and myosin, both contraction and relaxation are highly regulated processes, however, muscle relaxation requires highest levels of chemical energy.

Overcome the electromagnetic forces of attraction between the actin and myosin, *i.e.* enhancement of repulsion forces between them; It requires more chemical energy than the opposite events.

Furthermore, during contraction, *i.e.* the dimensional change is probably secondary to chemical energy variations in which electrical charges are neutralized or modified so fast in one or another address, being, under normal conditions; a fully reversible process, therefore, chemical energy needed is astonishing accurate and almost always available. The size of this functional units is independent of the colloidal particle size (1 to 1.5 x 10^6 g) into which myosin breaks up on extraction, thereby can be expected to be much smaller than this size.

Muscle Changes After Death, Rigor Mortis

Moreover, the contractile matter of muscle is built of functional units containing myosin, actin, titin, and nebulin. Since muscle contains no free ATP, it can be expected, therefore; that if the ATP concentration of muscle decreases, then, the number of contractile units also decreases proportionately. The ATP concentration of muscle decreases after the death of the specimen, and thereafter, rigor mortis appears; so, it is paradoxical that if the source of energy (ATP) diminish, the muscle is constantly contracted. Current method for the assessment of ATP are based in bio-luminescence reactions -luciferase and luciferin reactions-. When ATP molecule should be estimated, extracts of the tissue can be treated with acid-hydrolysis in a limited form, and the quantity of ATP is inferred from the detectable quantity of labile phosphate liberated. Muscle contains *in vivo* a not immensurable amount of free phosphate. ATP is not associated with matter of non-biological origin.

Experimentally, ATP content and the force or tension developed by muscle decrease in a proportional way but often inversely. Thus, a parallelism or positive correlation would support the theory of ATP as source of energy, while a lack of parallelism (negative correlation) would plead against it, favoring molecular hydrogen -from melanin and water- as source of chemical energy.

Molecular Hydrogen and Rigor Mortis

Muscle unavoidable changes after death, producing a constant contraction state called rigor mortis (stiffening of the corpse) due to biochemical changes in muscle that happen after death; no one has could explain it adequately in basis of glucose or ATP as chemical energy source.

During many years, it was thought that the rigor mortis (stiffening); could be due to coagulation of the muscle plasma either spontaneously or by a process akin to the coagulation of blood, whilst others proposed precipitation of the proteins by the lactic acid produced after death.

These old concepts arose from ancient theories (coagulation) proposed by Kühne in 1864, and Schipiloff in 1882 (lactic acid). The “alkaline rigor” (Hoet & Marks 1926) or third stage, during which no acid is produced at all, yet the stiffening occurred as rapidly and as completely as in normal rigor, was apparently explained by the discovery of Engelhardt & Ljubimova in 1939 about the ATPase activity of the structural protein myosin, that converts chemical energy into directed movement and can be viewed as a molecular motor. This protein comes in many shapes and sizes. Over 11 classes of myosin have been identified, and it is anticipated that more will be found as the search continues [9]. Often it has been recognized in one form or another in every eukaryotic cell examined.

Along decades, substantial effort has been dedicated to determining the physical-chemical basis of the energy conversion by myosin. All the myosin isoforms examined until today, exhibit similar kinetic strategies and share common features of the cycle that converts chemical energy into directed movement. Contrary to initial expectations but in accordance with our work, ATP hydrolysis in myosin is not coincident with the force-generating step (Rayment, 1999). Instead, ATP binding initially reduces the electrical- chemical affinity of myosin for actin, which depends of a fine tuned by billion years of evolution, balance of repulsive and attractive forces; after which hydrolysis of ATP occurs rapidly, due to are compounds with thermodynamic instability and kinetic stability and results in a metastable ternary complex between myosin, ADP, and inorganic phosphate (Pi). However, it is possible to consider this stage as the highly complex and much energy requiring relaxation state, begin to lose gradually chemical energy and a careful disorder ensues, whose first manifestation is by binding of ATP to myosin.

During this time (ATP hydrolysis) there is rapid energy exchange manifested by a fast interconversion between substrate and products, conceivable entropy of microenvironment is growing as level of chemical energy lowed, in significant part because ATP hydrolysis results in absorption of available chemical energy, and in presence of low levels of chemical energy, the astonishing accuracy in space, time and location of biochemical reactions tend to decay; therefore the fast interconversion between substrate and products is more a signal of incipient disorganization than higher chemical energy availability.

Theoretically, release of the hydrolysis products from myosin is catalyzed by the rebinding of myosin to actin, however, under our research experience, chemical energy levels are recovering, and order ensues. It is believed that the energy transduction step seems to occur during product release, but chemical energy from water dissociation is present all the time. The chemical energy that comes from melanin is astonishingly accurate, so slight changes in microenvironment tend to lower it. For example, the movement of sodium, potassium; and finally, by energy absorption by hydrolysis of ATP.

Thus, myosin is an unusual enzyme in that the chemical step occurs at a different point in the contractile cycle from the energy transduction event. This most likely arose because myosin spends a very small time attached to actin, even almost all biochemical reactions are in the Nano and Picoseconds scale.

The finding that ATP hydrolysis in myosin is not coincident with force-generating stage, presents several complex and interesting biochemical problems that can be resolved arising from water dissociation.

Question: How is the hydrolysis of ATP coupled to energy transduction?

Answer: Diminishing basal chemical energy levels that comes from water dissociation.

Question: how does myosin catalyze the hydrolysis of ATP?

Answer: Cascade of biochemical events during muscular contraction, starts with lowering of chemical energy levels (from water dissociation), and under normal conditions the very first step is the nervous impulse that induces a depolarization of membranes, which is enough to starts the careful disorganization of the highly complex processes of energy generation and distribution. One of the first manifestations of low level chemical energy is the ATP binding to myosin.

Question: what is the physical basis of the metastable state?

Answer: Low levels of chemical energy (from water dissociation).

The conformational changes induced by ATP hydrolysis and interaction with actin are better explained by variations in chemical energy levels coming from water dissociation.

Thereby, ATP is not a chemical energy currency, oppositely requires available chemical energy to be reconstituted every 20 seconds. Furthermore, we found that ATP hydrolysis absorbs (consumes) chemical energy, and ATP synthesis, from ADP; releases chemical energy. In spite there is expenditure of chemical energy to achieve the ATP synthesis, at the end the balance is positive (exergonic reaction).

The progressive contraction process that happens in rigor mortis can be explained in the following way: when generation and distribution of energy from water dissociation is stopped or substantially altered, the patient dies, thereby, chemical energy that comes mainly from the melanin became insufficient to sustain the highly accurate biochemical order that life needs.

Once water dissociation is stopped or substantially damaged, the degraded ATP, this is ADP, cannot be reconstituted to ATP every 20 seconds, as normally happens. Other significant effect of the low levels of molecular hydrogen is that chemical energy levels in muscle progressively decreases more and more. With low levels of chemical energy, the muscle tends to contract, and this loss of chemical energy from water dissociation, that is progressive, so muscle contraction too. Explanation is that repulsive forces (more energy) are then dominated by the forces of attraction (less energy) between the actin and myosin, and then the attractions between these protein complexes increasing gradually more and more, which is manifested by a sustained, progressive, and eventually irreversible muscle contracture; so-called rigor mortis. Muscle relaxation cannot be achieved given that is a highly complex process that requires highest levels of chemical energy, that unexpectedly are basal levels.

After death, the ATP content is gradually degraded post mortem, without re-constitution. Therefore; the amount of hydrolysable phosphate decreases while the amount of free phosphate increases, this is: a kind of negative correlation; and thus, even a slight error in the free phosphate estimation makes the results of the ATP estimation doubtful. Moreover, Borbiro and Szent-Gyorgyi, found that under experimental conditions, that muscular tension and amount of ATP run parallel by inverse, at least initially after death; but at the end of the third hour, paradoxically; the muscle develops no more tension and does not contract on thawing, and turns completely inelastic; a paradoxical finding if ATP would be energy currency. By other side, it is an expected finding that gives supporting to our finding that higher amount of chemical energy (from water energy) is required to relaxation.

The muscle at this "stiff" point still contains a not negligible amount of labile phosphate. Whether this labile phosphate (P) is derived from ATP or ADP could not be stated clearly, but supposedly ATP is the main source. If that labile P is derived from ATP, then this ATP must be, at least theoretically; different from the rest, because it is no longer split by the muscle or it is split exceedingly slowly and has no detectable influence on contractility or elasticity. Therefore, ATP activity is substantially determined by the energy that comes from the water dissociation. This "residual" hydrolysable phosphate (ATP or ADP) apparently with no activity, was found in strikingly similar proportion in all experiments reported by Prof. Szent-Gyorgyi. A fact that means that chemical changes induce by molecular hydrogen are subtle, and by other hand, the astonishing regularity by which melanin dissociates the water molecule, can be observed in different individuals; and is manifested by the regularity of their histochemical processes.

The relationship between ATP and muscle contraction is not purely linear, *i.e.* after two hours, the concentration of ATP seems to be independent of the degree of contraction. This fact is difficult to interpret unless you consider the possibility that the muscle contraction and relaxation are opposed facts that do not primarily depend on ATP levels, instead it is a secondary phenomenon in the cascade of chemical reactions that lead to muscle relaxation and contraction. An interesting point, is that decrease of ATP concentration is linear; which means that: the rate of its disappearance is independent of its concentration. The most likely interpretation of this striking unexpected fact was that the splitting of ATP depends on some change in neighbor molecules in the contractile matter itself. In our opinion, for instance, the Heme Group of Myoglobin can keep dissociative activity by significant periods even after death, and gradually stopped it completely in a certain period, that is variable in every case.

By other side, myosin seems to have ATPase enzymatic activity only when muscle is in contracted condition. In a quarterly part of their experiments, and unexpectedly, the ATP concentration did not fall at all during the first hour after death. Then, theoretically; muscle must be relaxed; however, rigor mortis appears.

This could be explained by the chemical energy that still can comes from melanin after death, even by brief time; or the chemical energy that preserve ATP concentration is derivate from myoglobin in form of molecular hydrogen (H_2).

The intrinsic property of melanin transforms light energy into chemical energy does not stop abruptly with the death of the person, at least by two things: melanin is the most stable substance man ever known, and that melanin can dissociate the water molecule both inside as outside the body or cell. Probably what is primarily altered after death is the distribution more than generation of power.

Differences Between Relaxation and Contraction

The physical and chemical differences between muscle contraction and relaxation are enormous, therefore, it is not rare that observed energy level differs completely from what they could expect, as the energy level of the relaxed muscle is significantly higher than those of the contracted muscle; a concept completely opposed to current dogmas.

Physical characteristics of muscles modifies substantively during normal activity, thus become hard when they contract, and being soft and plastic at rest. The muscle becomes, astonishingly; hard when it enters what we considered muscle activity. The difference is huge, and we take it as a natural matter of fact because we see it every day. Thereby, it means that there are vast differences in the physical-chemical constants of muscle tissue in harmony with the condition and circumstances of functional activity.

The chemical mechanisms of these changes, elucidated at least partially by Prof. Szent-Gyorgyi, start defining that actomyosin is the contractile matter, that is a protein molecular complex formed by actin and myosin, appearing that the other constituents of muscle tissue are at service of these two proteins. These proteins, during relaxation; are not joined together, just are close each other but are histologically dissociated as free actin and free myosin. The great affinity that these two proteins have, are kept apart in a very precise manner by electric repulsive forces present between actin and myosin, which are predominant during relaxation, but they seem to diminish during the contraction. In our experience, when chemical energy (from melanin and myoglobin in form of H_2) is available, or it is adequate, then repulsive forces between actin and myosin predominate and the muscle is in a relaxed condition. That means that relaxation is a highest energy state, and contraction is the lowest energy condition. This is the reason why, repulsive forces, so far; are considered complex and the role of ions and ATP are poorly understood. The presence of forces that repel the actin and myosin during relaxation indicates the presence of chemical energy, which when it drops, the repulsion force weakens and then the actin and myosin are attracted between them. On the other hand, the movement of ions and ATP levels also imply the necessary presence of chemical energy, as any chemical reaction involves the exchange of energy. We can summarize by saying that it is the same tissue that changes significantly depending on distinct levels of chemical energy. But the energy levels are completely opposite to what we were accustomed. There is widespread belief - erroneous - that the contracted muscle requires higher levels of energy, when it is the opposite.

Experimental data had been shown that ATP is adsorbed at specific union sites of myosin. The molecule is characterized by several deep clefts and pockets, one of which forms the nucleotide binding site and gives it a state of charge which repels actin; but contraction it is not initiated by ATP binding; instead it is started by a previous phenomenon that probably is membrane depolarization at neuromuscular junction. The movement of sodium and potassium is sufficient to increase the entropy of the system and the highly accurate generation and distribution of energy by melanin, is slightly diminished, but in sufficient amount so the next steps occur, *i.e.* nucleotide binding. Thereby, the energetic level decreases even more by effect of ATP, but in a very fine-tuned form, so the biochemical cascade of contraction, that implies a careful lowering of chemical energy levels, happens accurately again and again.

Thus, myosin or is an unusual enzyme in which the force chemical step occurs at a different point in the contractile cycle from the energy transduction event or ATP hydrolysis has different function. It is most probably the later, otherwise try to explain an unusual enzyme will be so difficult and lengthy as had been the fruitless searching of ATP energy release explanation after six or seven decades.

Myosin is physically a very soft, precise, accurate liquid gel and so is actin, and these physical-chemical characteristics are not reached by chance, given that reflects a highly ordered composition. When they (actin, titin, myosin and nebulin) unite orderly to form actomyosin - this union process also requires highly consistent chemical energy-, a new substance is formed which has new and different qualities, something that is usual in Nature.

The changes that Prof. Szent-Gyorgyi observed in a solution of actin and myosin were consistent with a significant increase in viscosity, which means dominance of attractive electrochemical forces. Moreover, if the two proteins, in enough high concentration, In the case in striated muscle where actin and myosin are present, characteristically; 8% myosin and 2% actin, on union they will form a stiff and hard gel, being a physical-chemical condition like muscle contraction period of lower chemical energy. If ATP is present and hydrolyzed, then the energetic level goes down even more, thereby, attractive electrochemical forces between actin and myosin keep and gain predominance, thereby the gel goes over into a new modification in which molecular distance between particles will be shorter due are closer. This is the typical muscular contraction.

Therefore, ATP hydrolysis does not mean higher levels of power, certainly the opposite fact. Chemically, ATP hydrolysis, to degrade to ADP absorbs energy (breakdown of bonds consumes energy). This is consistent with observations of the Professor Szent-Gyorgyi, but no with his interpretation, and are congruous with our work findings about that decreasing levels of chemical energy, then, the muscle tends to shrink, to contract. Prof. Szent-Gyorgyi wrote in those times, that muscle relaxation and contraction would be easy to explain if some biochemical process would exist that could be explain the origin of common levels of energy along the cell (Fig. **3**).

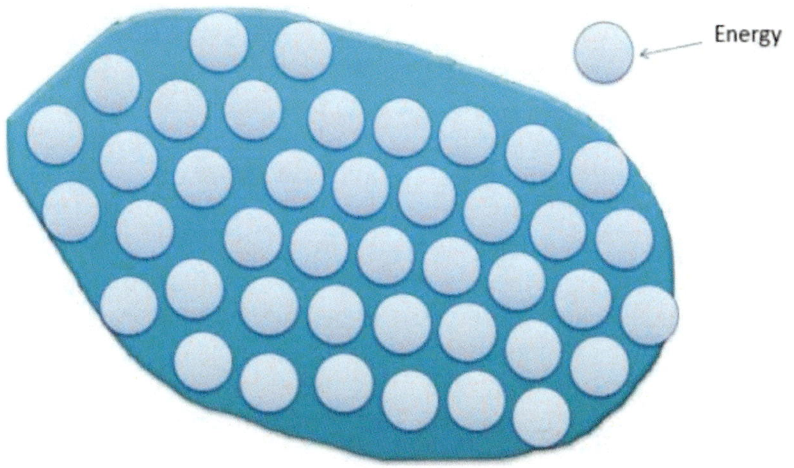

Fig. (3). Common energy levels could be represented as something that has a uniform distribution along inside of the cell. In this scheme, we drawn them as pale blue spheres of molecular hydrogen, transporting energy; distributed evenly on the inside of the cell. Theoretically, the common levels of energy, according to current theories, should be given by the ATP molecules, but on the one hand, neither Mitochondria nor ATPs are distributed evenly inside of the cell, and on the other hand, each molecule of ATP requires to be re - constituted every 20 seconds; so, try to maintain these common levels of energy implies an unsurmountable energy expenditure. However, if we talk about gas bubbles (of H_2) then is different due to melanin produces them incessantly, night and day, and once are released (by melanin), spontaneously tend to be distributed uniformly inside the cell, following Simple Diffusion Laws. Prof Szent-Gyorgyi couldn't go beyond, *i.e.* extrapolated his knowledge to the clinic, since in the first half of the 20th century not be understanding as eukaryotic cell absorbed energy from the environment. It was not until the second half of the 20th century that theory chemo-osmotic of Mitchell made his appearance, but anyway it was not very useful. Before continuing with the extraordinary work of Prof. Szent-Gyorgyi we would point out that when the ATP is transformed into ADP the energy is absorbed, something that is opposed to general belief; therefore, when ATP is degraded, resting muscle energy level decreases, and the actin and myosin molecules approach. Keep separate the actin and myosin, *i.e.* requires energy, which certainly is provided by melanin and not for glucose. While most decreases the energy level of the muscle, - by the degradation of ATP and thereby more energy is absorbed - more it shrinks by attractive electromagnetic forces between actin and myosin.

During the cascade of biochemical and physical processes by which a muscle can contracted, the ATP also is hydrolyzed, thereby absorbing (consumes) energy which backing attractive forces, as the need of motions demands. But ATP is also hydrolyzed, though a much slower rate, therefore, in a separate way, in resting muscle, because muscle relaxation is a highly complex process, therefore, ATP is slowly degraded and thereby chemical energy decreases also in a slowly and localized form, that is not followed by a complete contracture, furthermore, there is mild, moderated, unperceived contraction that maintains a relaxation, and at the same time, a proper muscle function and shape. This would explain, at least in part, what we call muscle tone.

Theoretically if oxygen is present, hydrolyzed ATP would be reconstituted and its concentration is kept at adequate level, however, it is not only oxygen what ATP synthesis need, also requires available chemical energy in form of H_2; and when the dissociation of water increases, the levels of hydrogen (H_2) and oxygen (O_2) molecular will recover normal levels because of the breakdown of the water molecule; and so far, the levels that are detected relatively easily are those of oxygen. Even, paradoxically, during muscle contraction the blood vessels are constrained and thereby there is no blood-flow so there is no oxygen and theoretically hydrolyzed ATP cannot be reconstituted due comprised blood flow. However, this happens especially after death, in which blood flow disappears; but not in a live patient. The consequence of this significant degradation or hydrolysis of ATP will be that the repulsive forces between actin and myosin diminish when ATP is transformed to ADP and chemical energy is consumed, as happens in any chemical reaction; and the two proteins get together forming the stiff and hard actomyosin. There being no enough ATP present to be hydrolyzed and thereafter absorbs accurately available chemical energy, repulsive forces will predominate perennially and thus actomyosin will not relax but remains stiff and hard due to available chemical energy is not sufficient to bring the muscle to a normal relaxation stage. This condition is known as rigor mortis, which has been difficult to explain on basis of the hydrolyzed ATP and glucose as energy source, but it is different when molecular hydrogen from water dissociation is included.

The thorough description of the Prof. Szent-Gyorgyi, who said that muscle relaxation and contraction would be simple to explain if a common energy source exists, is consistent with our finding that the common energy comes from the constant, incessant breaking and re- forming of the water molecule by melanin, and also in significant and accurate part by myoglobin, so she also have the capacity to dissociate irreversible the water molecule, and on the other hand, the fact that when the ADP is transformed into ATP energy is released, but when it goes from ATP to ADP energy is absorbed.

Prof. Szent-Gyorgyi rightly wrote that there must be oxygen and blood circulation so that ATP is reformed. Although we interpreted it in a different manner, because when there are normal levels of oxygen and blood circulation is efficient, then the levels of chemical energy that emanate from the melanin and myoglobin are suitable, these facts indicate the bio-energetic system is working. When there is no available oxygen in tissues or blood circulation stopped or sluggish, we can think that melanin has stopped transform light, visible and invisible into chemical energy, and when this happens, the failure is generalized, and death ensues within minutes.

The actin filaments are built of so small globular molecules, and the electrochemical forces that holding up the complex actin globules, which together normally form filaments, whose presence is important to contraction and relaxation processes; due to release of the hydrolysis products from myosin is catalyzed by the rebinding of myosin to actin. The energy transduction step occurs during product release but without a doubt are mainly the result of the diatomic hydrogen and the high-energy electrons that emanate continually from melanin, day and night.

Therefore, the relaxed muscle state is a complex and high-energy steady state, and oppositely, the contracted stage is the lowest- energy stage, thereby, contraction can be considered more as a lower energy process.

Muscle is not an anatomically homogeneous tissue. There are distinct kinds of muscle (*i.e.* smooth, heart and cross-striated muscle), furthermore, there are significant differences between the various muscles of the same sort within the same animal. There is substantive difference in histology between the various body muscles, however, at molecular levels, differences almost disappear. In one muscle the fibers are parallel, while in others they follow a more complicated course, making evaluation of energy relations difficult, *i.e.* ciliary muscle of the eye. There is considerable difference, also, in the macroscopic composition of various muscles. However, the contractile matter, actomyosin, at molecular level, has surprisingly slight differences, even with other species, *i.e.* in its relaxed condition, actomyosin is a soft gel which could easily be damaged even by soft mechanical injury were it not protected by surrounding tissues as connective material, fasciae, collagen fibers and a sarcolemma, and this; in all species.

Per the theory outlined by Prof. Szent-Gyorgyi, contraction is more like a spontaneous process going together with an increase in the chemical energy transported by ATP which is released when ATP is hydrolyzed. However, paradoxically; contraction results by a drop of available chemical energy. Thus, contraction theoretically should occur spontaneously, wherever the ATP-actomyosin system is present in a suitable ionic milieu, that allows accurate variations in chemical energy levels; and as expected, the system should persist in the low- energy contracted state. Due to Prof. Szent- Gyorgyi did not known the energy that comes from melanin, his conclusions were based in hydrolysis of ATP. The energy carried by molecular hydrogen is accurate, so when it was absent or low, the biology of the muscle system cannot reach the muscle relaxation stage, which is the highest energy level.

In the intact resting muscle, however, in spite we find ATP molecule in a complete form, linked lousy to actomyosin, but still the system does not contract; thereby contraction seems as is being inhibited by some unknown mechanism, that we know now is the basal levels of chemical energy provided by melanin and myoglobin. If we want the muscle to go over into a contracted state, we must abolish this inhibition inducing a drop in chemical energy level, *i.e.*, by ATP hydrolysis. In our experience, this inhibitory mechanism comes from common energy levels arising from melanin and myoglobin.

In experiments performed at the U.S., in the marine Biological Laboratory in Woods Hole, in 1949, Dr. Szent-Gyorgyi permeabilized the muscle membrane with glycerol, added ATP, and watched the muscle contract, thus providing a proof that the contractile system consists (under his experimental conditions) of, at least; actin, myosin, ATP and glycerol [10]. Although he believed that ATP provides the energy for the contraction, which not is so, because what happens in any molecule in which is break their links is consume it.

In our experience, glycerol, applied to cell membrane, induces an irreversible disorganization, due its effect in the generation and distribution of energy, that is hampered due to are astonishing accurate processes. This is the reason why the muscle contracted by decreasing energy levels that are required to dominate the repulsive forces between actin and myosin, favoring the forces of attraction and therefore the muscle contracted.

When we voluntarily contract a muscle, what we are doing is to decrease the levels of chemical energy "voluntarily", which favors the electromagnetic forces of attraction between actin and myosin. When we voluntarily relax a muscle, we are "voluntarily" elevating the energy levels of that tissue, which favors the repulsive electromagnetic forces between actin and myosin.

At light of the unraveling of the intrinsic property of melanin transforms visible and invisible light into chemical energy through the water molecule breaking, as chlorophyll in plants; the analysis of the results, but not the conclusions, of Dr. Szent-Gyorgyi are consistent with our findings that the ATP, to become ADP absorbs (consumes) energy diminishing the quite accurate levels of chemical energy. Thereby, the real ATP function is diminishing muscular chemical energy level and also temperature, in a very precise manner, this is: whilst more ATP is downgraded to ADP, there will be more muscular contraction and *vice versa*, so, when chemical energy levels increase (ADP to ATP releases energy), then an accurate relaxation ensues. It seems that the depolarization of the muscle membrane starts the descent of the energy levels and therefore the muscle contraction begins, while the ATP only complemented in a fine-tuned way the event. Summarizing: the muscle tissue in relaxed state requires higher chemical energy levels than the muscle contraction. In other words: contraction requires less energy than relaxation.

The energy that comes from the melanin in the form of molecular hydrogen and high energy electrons, it is released in a symmetrical way, in all directions, constantly, so consistent; very accurate, forming what might be called common or basal levels of energy, specifically available chemical energy, this is: wearable energy. A power supply common that Dr. Szent-Gyorgyi postulated as an explanation to the particularities of muscle contraction, but that could not transfer to clinical practice because he did not find it or did not know at that time the unsuspected intrinsic property of melanin to split the water molecule, thereby he believed that ATP was the explanation.

Furthermore: when the chemical energy that melanin constantly releases is suitable, the system is at its maximum power level, therefore, muscle is relaxed; or at its entropic lower, when the ATP is hydrolyzed to ADP absorbs (consumes) energy, and therefore common or basal muscular energy level decreases, what increases entropy of the molecules that make up the contractile substance or actomyosin, these changes its conformation, and contraction occurs.

In other words: when available chemical energy levels decreased, attractions are optimized, and repulsions are diminished. It is expected that poisons that expectedly disorganize generation and distribution of chemical energy from melanin, like caffeine, quinine, mono-iodine-acetic acid, chloroform and other anesthetic agents, it is already known that produce contracture, overall when are administered systemically; meanwhile, when are applied locally, developed muscular tensions are small, which means that only a small fraction of the contractile substance is at any time in the contracted state.

The explanation is that the poisons above cited, decrease the transformation of sunshine into chemical energy in several body systems, and result are poor under experimental local conditions, because in the live subject, muscular chemical energy levels are turned down but in a generalized or systemic way. It is something similar that we have seen in other biological or non-biological systems, as the lowering of the main chemical energy of a system is simpler, more repeatable, more attainable, cheaper and faster than try to raise it. Seems similar as with Calcium chloride, when injected into the muscle fiber, even in minimal quantity, causes an immediate local contraction [11], explainable by the great affinity of melanin for calcium, about one thousand times more than bone.

Turning to the work of Prof. Szent-Gyorgyi, he described that a satisfactory method of abolishing those unknown inhibitions is through freezing with subsequent thawing, now we explained it by the fact that cold down-turn the generation and distribution of energy both in plants, animals (hibernation) and humans; being a method that also has the advantage that the muscle can be kept in the frozen state, packed in dry ice, for days with undiminished contractility, this is: in presence of low level of chemical energy, the muscle should remain contracted. Because the work carried out by the electromagnetic forces of attraction does not imply a constant expenditure of energy on the part of the organism.

On thawing, the frozen muscle, if containing the physiological amounts of ATP, contracts rapidly and develops maximal tension but not because the ATP provides the energy needed for contraction, but because the system is completed, and the biochemical cascade events occur as it should be. Low temperature decreases turnover rate of water dissociation in all living things. The contraction, elicited by the freezing and subsequent thawing, is due to that generation and distribution of energy that come from melanin and myoglobin, usually is substantially affected, diminishing it and therefore a contracted state or period ensues. It is developed relatively slowly while the sudden change in temperature may act as a fast down-turn energy factor and elicit a fast contraction.

For us, the explanation is that cold reduces importantly the transformation of light into chemical energy both in humans, animals (hibernation) and plants, therefore increasing temperature, chemical energy levels begin to rise gradually, and in that case, relaxed state is obtained.

The interesting and careful experiments of Prof. Szent-Gyorgyi were performed usually in Ringer solution containing 0.001 M MgCl. In all experiments Prof. used glass-distilled water because of the deleterious action of copper that is present in common distilled water. It is already known that metals significantly decrease the transformation of light into chemical energy, probably due to melanin has a high affinity for metals and the physical- chemical changes in the water itself that metal induces.

Unloaded fibers (with no weight) will contract at room temperature usually to one-fourth or one-fifth of their full length at rest. The change in the length of the muscular fiber is notoriously exact, and a displacement so precise, so delicate, is better explained by attractions and electromagnetic repulsions at microscopic level, then by the supposed energy that releases the ATP, which to date is theoretical, because you do not even know how this energy is released, this is: in the form of heat? In the form of a thrust? *etc.*

In clinical practice, we can observe only the fact that the levels of chemical energy, when decrease, the muscle contracts. For example, in athletes, after receiving a blow, the generation and distribution of energy of the affected tissue decreases, resulting in a painful contraction (cramps), so intense. Something similar occurs in the metabolic problems - diabetes-where the appearance of night cramps, is common which is paradoxical, because if glucose levels are high, theoretically, the patient and the muscle have much energy.

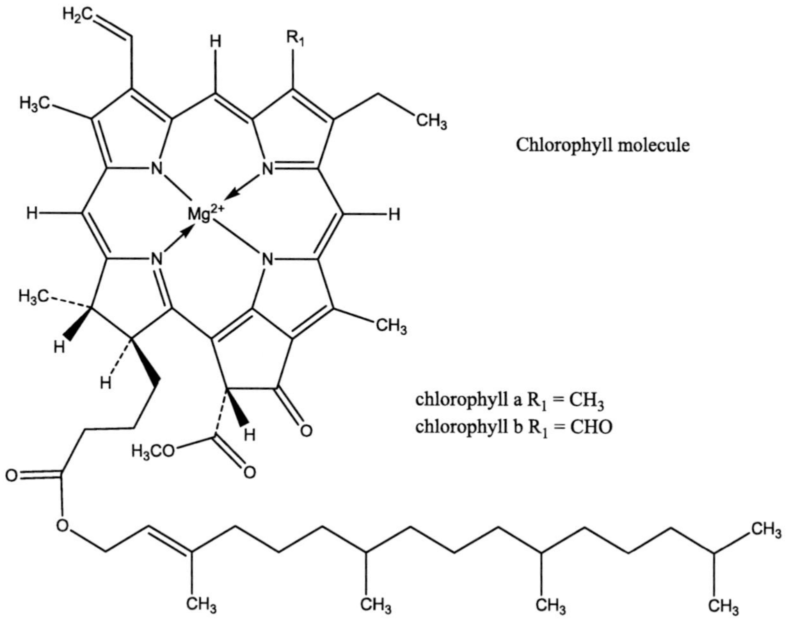

Chemical Formula: $C_{55}H_{71}MgN_4O_5^{2+}$
Exact Mass: 891.53
Molecular Weight: 892.48
m/z: 445.76 (100.0%), 446.26 (73.2%), 446.76 (41.1%), 447.26 (11.2%), 447.27 (4.2%), 447.76 (2.9%), 446.26 (1.5%), 447.77 (1.2%)
Elemental Analysis: C, 74.02; H, 8.02; Mg, 2.72; N, 6.28; O, 8.96

Fig. (4). The Heme group of the chlorophyll molecule.

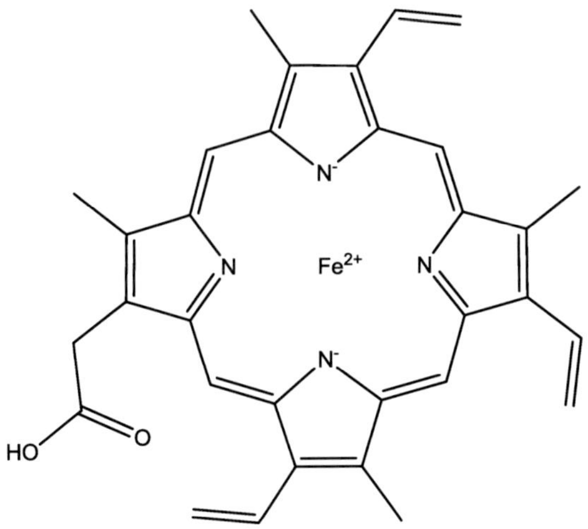

Heme group
(Hemoglobin)

Chemical Formula: $C_{32}H_{28}FeN_4O_2$
Exact Mass: 556.16
Molecular Weight: 556.44
m/z: 556.16 (100.0%), 557.16 (37.3%), 558.16 (7.6%), 554.16 (6.4%), 555.16 (2.3%), 557.15 (1.5%)
Elemental Analysis: C, 69.07; H, 5.07; Fe, 10.04; N, 10.07; O, 5.75

Fig. (5). The Heme group of the hemoglobin.

Bioenergetics and Heart Muscle

The most basic property of the heart is that it is a muscle, and the main property of muscle is that we cannot understand it [12]. The more we know about it, the less we understand, and it looks as if we would soon know everything and understand nothing. We see it too much, so we don´t look it.

The situation is very similar in most other biological processes and even in pathological conditions, such as the chronic-degenerative diseases and chronic metabolic disorders, for instance glucose has been drive by body since the beginning of times. This suggests that some very basic information is missing and the unsuspected role of melanin as bio-energetic molecule can be very helpful to find the answer.

In his masterly description of the energetics of myosin, Prof. Szent-Gyorgyi wrote that muscle´s main contractile protein is myosin. It converts – in theory- the chemical energy of adenosine triphosphate (ATP) into motion. However, the first disturbance was caused by Gergely, Perry, and Mihaly, who found that the complex structure of myosin is built by two kinds of subunits [13]: meromyosins;

that were subsequently isolated by Szent-Gyorgyi. The one that can sedimenting faster was called H (heavy) and another one denominated: L (light). There were twice as many L´s as H´s, and they were arranged in series, in a row.

What was disturbing about this finding was the fact that the L seemed to be involved in contraction, while the H alone interacted with ATP. How could the energy have released on the H support work done by the L´s?

In our opinion the explanation is ATP absorbs (consumes) energy to become ADP, thereby domain chemical levels of energy are turning down in the immediate surroundings. Thereby a change in chemical energy levels in meromyosins H reach the meromyosins L. Furthermore, arrangement in series, in a row seems; as best geometry to optimize the generation and distribution of energy. Recall that energy is released symmetrically, in all directions, as growing spheres.

How a Bond-energy Could Lock Up in a Chemical Link Produce Work Somewhere Else?

In our experience, chemical energy levels, transported by molecular hydrogen offer a good explanation. When ATP hydrolysis interact with meromyosins H, energy is absorbed, then chemical energy levels are down-turned not only in the meromyosins H, but the process spread enough to reach nearby meromyosins L.

Other theories about began to look rather artificial. A new difficulty arose when Szent-Gyorgyi and Borbiro [14] made the discovery that the meromyosins themselves are built of a substantial number of much smaller subunits, tropomyosin's, which are held together only by secondary links, such H-bonds, van der Waals forces, and electric attractions, this is: no covalent bonds. The myosin "molecule" was thus no molecule at all, if we call a molecule a structure of atoms held together by covalent links.

Secondary links have not fixed valence angles, therefore have a great pliability to the structure. It became difficult to see how such a structure could "fold" and "unfold". It seemed more probable that contraction consisted of a rearrangement of tropomyosin's, a process that also depends of available chemical energy from melanin and myosin; which went into a more compact heap.

There must be strong attractive force between tropomyosin's to enable the myosin particle to withstand strain, and these forces must tend to pull the tropomyosin's closer together. The force of muscular contraction could thus be due to these attractive forces, in which case we would need force to stretch the particle out again, in other words, raise chemical energy levels again to basal stage. But it is simple if we consider molecular hydrogen produced through the dissociation of water molecule by melanin and myoglobin, that happens all the time, in a very precise and consistent form, night and day, and occurs inside the melanin molecule, due to its unique ability to absorb almost all wavelengths, that is from gamma rays to radio waves. Myoglobin is also able to dissociate water molecule, but under most limited conditions in comparison with melanin. Therefore, the energy that emanates from melanin molecule, in the form of molecular hydrogen and electrons of high energy, maintain a surprising consistent energy levels, therefore most of biochemical body processes are based on decreasing the level of basic energy, rather than increase it (Fig. **6**).

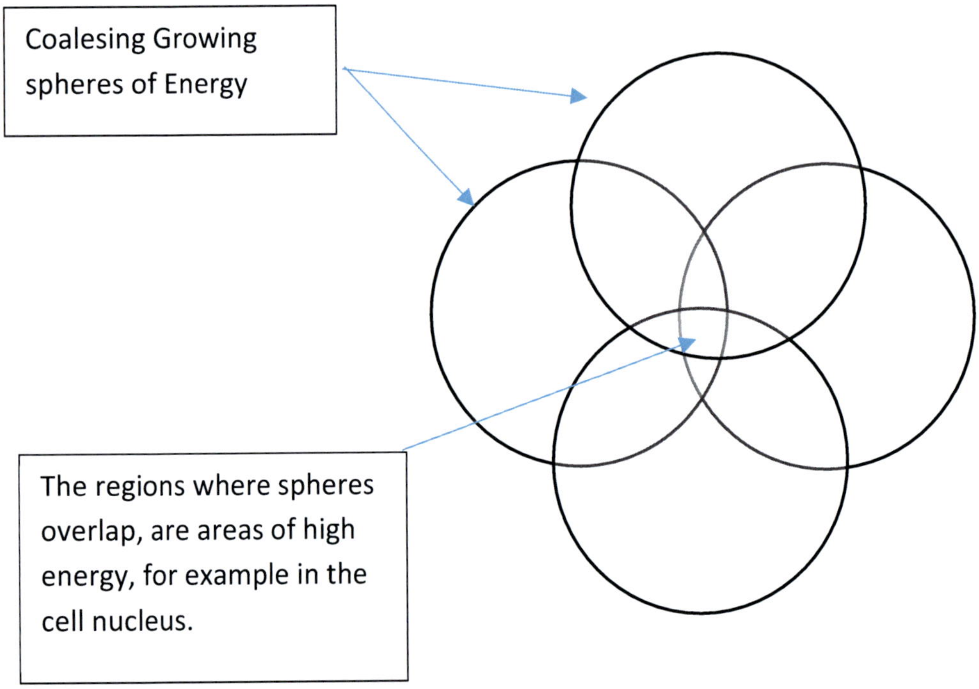

Fig. (6). The regions where spheres of energy overlap, are areas of high energy, for example in the cell nucleus, in the midbrain, around substantia nigra, *etc*.

Therefore, our theory of contraction is simple and attractive. The long-lasting difficulty with ATP hydrolysis function, which seems to induce contraction and foots the energy bill, now is overcome. Thereby, we are in full accordance with Prof. Szent-Gyorgyi, who explain the role of ATP as energy exchange universal currency complicates theory piping, because on one hand, basic or common chemical energy comes from melanin and myoglobin, and on the other hand, when the hydrolyzed ATP becomes ADP energy is absorbed (consumed), not freed. So, we can discard the elaborate and numerous theories that try to explain how, in the case of the tropomyosin's, the hydrolysis of ATP molecule can influence links weak as energy is absorbed or consumed; and how is work for high distance to the site of action of ATP.

Prof. Szent-Gyorgyi set too accurately that it is not actomyosin itself which is essentially different, but the machinery which provides actomyosin with chemical energy, in those times the dogma was ATP hydrolysis, but as gradually we understand the relevant role of melanin and myoglobin as bio-energetics molecules, then the function of glucose bonds and ATP hydrolysis as sources of energy, emerge as the phlogiston of XX century.

High Energy Electrons

It is very interesting Prof. Szent-Gyorgyi reflection about that, as rule; no electron can stay in the excited state longer than 10^{-8} second, and if the molecule is unable to dissipate its excess of energy, the electron will drop back, shooting out its excess of energy in 10^{-8} seconds at most.

The electron moves close to light speed; therefore, the electrons are thought to be transferred to the respiratory chain because of incidental encounters, which are explained in terms of vibrations and rotatory motions of the parameters involved that go as far as the quantum-mechanical tunneling of membrane barriers.

Contrary to the random and chaotic organization of the mitochondrial energy transfer assumed in the classic corpuscular model, functional movements and changes in the cell take place in a highly ordered way. Otherwise, the feedback control system cell and the entire human organism would not be able to exist in the end.

This means that a molecule is fluorescent, indicating that the molecule can accept energy without dissipating it. Trying to deepen in as it is possible that high energy electrons have a role in biology in less than 10^{-8} seconds, we will refer to the atom consists of a nucleus surrounded by a system of electrons. By sharing one or more electrons, atoms can join to form molecules. In such a molecule, as a rule, every electron belongs to one or two atoms.

However, the study of crystals and metals has revealed the existence of a different states of matter. If a considerable number of atoms is arranged with regularity near, *e.g.* in a crystal lattice, the terms of the single valence electrons may fuse into common bands. The electrons in this band cease to belong to one or two atoms only, and belong to the complete system. These bands or energy levels are separated from possibly higher levels by forbidden zones. Under ordinary conditions all electrons are within the lowest band. If this lowest band contains the maximum number of electrons (2n if the number of atoms is n), as is the case with insulators, the electrons will be unable to transport energy. If, however, one of these electrons is raised by the absorption of energy to a higher level, and comes to be in what we call an excited state or high energy electrons, where it will move and transports its energy easily, and it will be impossible to say which is the atom to which the excited electrons belong, and the entire system can be looked upon as activated. By falling back to the lower level, the electron will give off its excess energy and perform work in a place distant from that of the absorption of energy. Thus, the absorption and emission of energy will proceed independently at separate places. It is probably that this state of matter, *i.e.* common energy levels, exists also in living system [15]. Protein molecules are systems built up of a considerable number of atoms, closely packed with great regularity.

The first indication of the existence of such common energy levels was given by the study of photosynthesis in plants. Emerson and Arnold [15] found that 2500 chlorophyll molecules form one functional unit. Warburg and Nagelian showed that four quanta of light are necessary to reduction of one CO_2 molecule. There are observations to indicate that these four quanta must reach the CO_2 molecule simultaneously. Gaffron and Wohl could calculate how many chlorophyll molecules must interact to absorb four quanta of extremes of visible light (violet and red), simultaneously at the weakest optimal illumination. Their calculation showed that only one thousand molecules can do this. These observations indicate that the electrons, raised to a higher energy level by the absorbed light, can move and transport their energy freely through the system of chlorophyll molecules (Fig. **4**), in a so far, poorly understood way.

The Visionary Concept of Common Energy Levels of Prof. Szent-Gyorgyi

Myosin is the main contractile element in muscles, it is a protein built up of fibrillar molecules. These molecules are arranged in small, primitive anatomic bundles. A considerable number of such primitive bundles form one microscopic fibril. So far, it is considered that the energy of muscular contraction is derived from the hydrolysis of adenosine triphosphate (ATP), a dogma recently demonstrated as wrong, due to ATP hydrolysis is not coincident with force generation phase. Thereby, the adenosine- triphosphate activity is bound up with myosin, but our measurements indicate that only very small fraction of myosin molecules can be endowed with such activity. However, in accordance with chaos theory; a minor change in the initial conditions can lead to very large changes in the subsequent evolution of the system. Thereby, the ancient problem, that is now resolved; about how the energy liberated or absorbed by a single molecule can be communicated to a vast number of similar molecules.

And the response seems logical: when the ATP is hydrolyzed to ADP, the energy is absorbed (consumed), therefore, the total chemical energy of the system decreases; which causes significant conformational changes in affected molecules, and in the case of the muscle, in the contractile elements the attractions are optimized, and the repulsive forces are minimized, with what appears the contraction, and *vice versa* in the case of relaxation.

In accordance with Prof. Szent-Gyorgyi, the concept of common energy levels gives an easy, adequate answer, but arising from water dissociation from melanin and heme groups of hemoglobin (Fig. **5**) and myoglobin; but not from glucose and ATP hydrolysis, as Prof. Szent-Gyorgyi believed.

Lactic Fermentation

The enzymes involved in lactic fermentation of muscle are hydro-soluble, this is: uses H_2 mainly; while the enzymes involved in oxidation are insoluble, *i.e.*, bound to the insoluble fibrillar proteins of the cell, which suggests that use mainly high energy electrons as source of energy. This difference can be explained if we suppose that the former is part of a system with common energy levels given by molecular hydrogen (from water dissociation). In lactic fermentation, no such common levels are necessary, for the single enzymes do not interact but react in series with soluble molecules, in that case, molecular hydrogen (H_2).

In our experience, cell soluble parts use the diatomic hydrogen and the insoluble ones uses mainly high energy electrons, both coming from melanin. By other side, in part of the oxidation system, the electrons wander from enzyme to enzyme, but close to light speed. The enzymes, being insoluble, have no significant free molecular motion and must be arranged so that their small reactive groups are at atomic distance, minimizing unexpected collisions. It is possible to arrange two large protein molecules in such way, but it is geometrically impossible to so arrange a whole series. Even, if we could device such an arrangement, it would still be incomprehensible how the energy liberated by the passing of an electron from one substance to the other, could do anything useful.

If the cell and with it the energy levels are disturbed in some way -generation and distribution of energy are astonishing accurate processes-, we can expect the electrons to fall freely to lower levels at random. This might explain why catabolic processes prevail over anabolic ones in damaged tissues (cancer?), why certain oxidations (catechol oxidase) are activated by damage, why chloroplasts refuse to build up carbohydrates, and why viruses refuse to multiply outside the cell.

The atomic structure is only the backbone that express these common energy levels. Possibly the cell wall is the border line of the common energy levels. It may be that the knowledge of common energy levels from melanin, this is: diatomic hydrogen and high energy electrons, will start a new period in biochemistry, taking this science into the realm of quantum-mechanics.

The intrinsic chemistry of melanin to transform electromagnetic radiations into chemical energy through the splitting and re-forming of the water molecule, fits very well the visionary concept of common energy levels proposed long time ago, but not explained so far; by Prof. Szent-Gyorgyi.

ATP and Energy

Supposedly, the energy needed to produce ATP is derived from the energy that foodstuffs contain in bonds, and once it is extracted, theoretically can be invested into the synthesis of the ATP molecule by two entirely different processes, being one oxidation, the second way is fermentation. In oxidation, the foodstuff molecule, for example sugar, is burned –combined with oxygen- to water and CO_2; in fermentation, it is merely decomposed to give lactic acid. Lactic acid (Fig. **7**) contains somewhat less energy than the sugar (Fig. **8**) from which it was derived, and this energy-difference can be used -theoretically- to build ATP and other biomolecules.

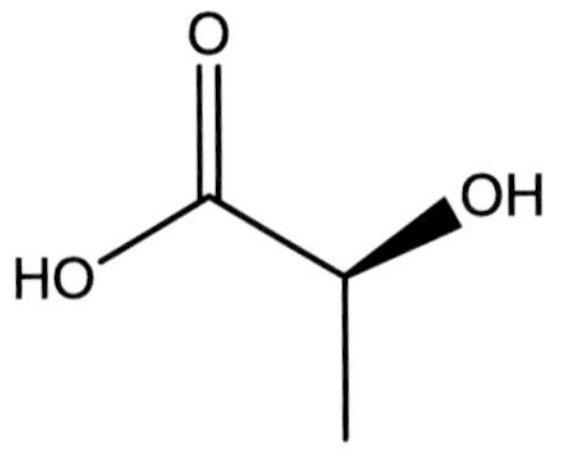

Chemical Formula:
$C_3H_6O_3$
Exact Mass: 90.03
Molecular Weight: 90.08
m/z: 90.03 (100.0%),
91.04 (3.4%)
Elemental Analysis: C,
40.00; H, 6.71; O, 53.29

Fig. (7). Lactic Acid.

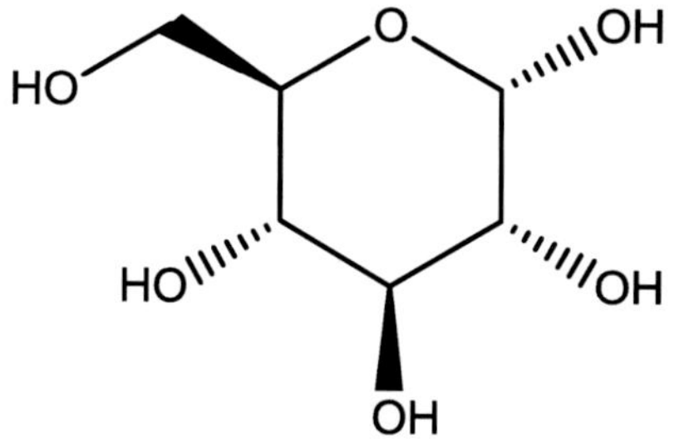

Chemical Formula: $C_6H_{12}O_6$
Exact Mass: 180.06
Molecular Weight: 180.16
m/z: 180.06 (100.0%), 181.07 (6.9%), 182.07 (1.4%)
Elemental Analysis: C, 40.00; H, 6.71; O, 53.29

Fig. (8). Glucose.

The difference in energy between lactic acid and sugar is not so great and therefore fermentation is a rather wasteful process. To render it more economical, oxidation must be coupled to fermentation, which oxidizes a small part of the lactic acid and, with the energy thus gained, converts the rest of the lactic acid into sugar again. However, evolutionarily has not sense due to oxidation demands a rather bulky mechanism whose development is expensive in energy and time of evolution, thereby cannot be explained easily.

It also seems that the solid state is essential for oxidation. It is not a very efficient but rather slow process which depends on available chemical energy and the oxygen supply. Fermentation can be affected by a number (a dozen or so) of soluble enzymes, thus it demands no solid structure or bulky molecular machinery. This is much the simpler process. It is true that eventually it also depends on oxidation but if time is given, a relatively poor oxidative apparatus will be sufficient to clear away the lactic acid produced. To all this we should add that fermentation can work very fast. Without oxidation, it could not work for very long because lactic acid would be accumulated, but for a short and rapid production of ATP is more suited than oxidation.

But not must forget that explanations have been implemented trying to of explain the way in that the cell extracts the energy of the links of glucose, which not is possible so the cell handles the glucose or the acid lactic as sources of biomass only.

During his work, the highest ATP concentration Prof. Szent-Gyorgyi has ever found was in the turkey breast muscle, 5 mg per gram. He theorized that added to this, there was a powerful system of fermentation which can rapidly replace the phosphate of the ATP used up. The system of oxidation, however, is very poorly developed, fact that is congruous with our work about oxidation is useless in biological systems, because molecular hydrogen is the important.

Consequently, the turkey breast muscle will only be able to work for a brief time but will be able to do it violently. The wing will be unfitted for a long-distance flight for which it is not intended. Once the animal has reached its goal, the poor oxidative mechanism will have time to clear away the lactic acid formed. Fermentation does not need structure, so is being performed by soluble proteins, thus the contractile matter is not embedded in the apparently solid structure of oxidative mechanisms. This results in the muscle having no color; it will be "white," since actomyosin is colorless.

We have retained substantive parts of Prof Szent-Gyorgyi text due the great clarity of its basic concepts, in spite his conclusions were wrong, due to Prof Szent-Gyorgyi tried to explain (fruitless) muscle energy from ATP hydrolysis. Per our work, when the ATP becomes ADP energy is absorbed or consumed, but not released; thereby, the muscle contraction is a decreasing in the levels of basal chemical energy that are given by physical and chemical properties of molecular hydrogen, and can be conceptualized as common energy levels, furthermore, it is a recent surprising finding that a relaxed muscle requires higher energy levels. Therefore, contraction required the careful decreasing of chemical energy levels, which are achieved relatively simply by hydrolysis of ATP to ADP, reaction that normally happens after membrane depolarization initially at neuro-muscle junction, a process that can be considered as part of the careful way in which our body modulate energy levels. And on the other hand, and paradoxically at light of current theories; when ADP is transformed into ATP, the energy is released, thereby, energy is provided to the surroundings during ATP reconstitution which should result in an increase of chemical energy levels, favoring the quite exact relaxation of muscular fiber. Our findings are consistent with the past observations of the Prof. Szent-Gyorgyi, but explanation is completely opposed; as energy levels rise and the reforming of the ATP is achieved -releasing energy-, the muscle relaxes.

Either way, the basic energy, common, or fundamental is those derived from the transduction of the electromagnetic radiations to chemical energy, this is: splitting water. The energy which theoretically is obtained from the ATP/ADP cycle is minimal, having too a substantive role in the regulation of the levels of phosphate compounds, which are compounds characterized by thermodynamic instability and kinetic stability, that must be regulated, otherwise, higher levels are toxic for the cell.

A rich store of ATP is of no use in a long run and fermentation would lead to an accumulation of lactic acid which would paralyze the muscle. Theoretically, the only system which can provide the energy in a protracted fashion is oxidation, which can keep pace with the energy-demand if there is an adequately powerful oxidative mechanism present, however, facts contradicted theory, at least in turkey breast muscle. However, this point of view, evolutivelly does not make sense. As mentioned above, oxidation is linked to the solid state and so in those animals which use their muscle for prolonged periods. In this case, the role of ATP is the same case, the decrease of the complete system energy level causes the muscle to contract. Remember that when the ATP is hydrolyzed, becomes ADP and energy is absorbed or consumed, but when the ADP is transformed into ATP energy is released. Changes in the level of energy due to the ATP and ADP probably not represent beyond 5% of the total and are merely complementary of the main source of energy which is in fact melanin and heme groups.

The whole oxidation depends on the oxygen supply besides protein complexes that make the work. Furthermore, the oxygen supply is not quite continuous, since the contraction of the muscle partially shuts down the capillaries. The oxygen supply is thus intermittent, while theoretically the metabolic oxygen demand is continuous. Therefore, to bridge the gap, and theoretically; there is a buffer between the two, a small storage of oxygen. The substance theoretically capable of storing oxygen is a dyestuff, closely related to hemoglobin (Fig. **5**). This "myoglobin" is rather dark in color. In species, which should stay under water for extended periods, the quantity of this myoglobin is especially large. But myoglobin cannot store oxygen, can produce it at her own, through dissociation of water molecule.

The dark meat with high myoglobin content is more elastic and rubbery. However, we found that Heme Group of myoglobin can split irreversible the water molecule, generating molecular both oxygen and hydrogen, like chlorophyll (Fig. **4**). Therefore, the Heme Group of myoglobin can produce oxygen as its own, more than stored it. By other hand, the physic-chemical attributes of molecular hydrogen explain the trophic effect on tissue.

Melanin, the Unsuspected Bioenergetic Molecule

Our discovery about the unknown intrinsic chemistry of the melanin to transform light visible and invisible, into energy chemical through dissociation of the water molecule, occurred in almost unexpectedly during an observational, descriptive study; initiated in 1990, about the three main causes of blindness in the world:

macular degeneration related to age, diabetic retinopathy and glaucoma. In the studio, the working hypothesis was to characterize in more detailed manner, the morphological particularities of the tiny vessels of the optic nerve (Fig. **2**), a delicate structure which, in humans is the equivalent of twelve human hairs altogether, and therefore important magnifications of this ocular element was required to carry it out observations, but these magnifications melanin began to be noticed. Normal Optic Nerve (Fig. **9**).

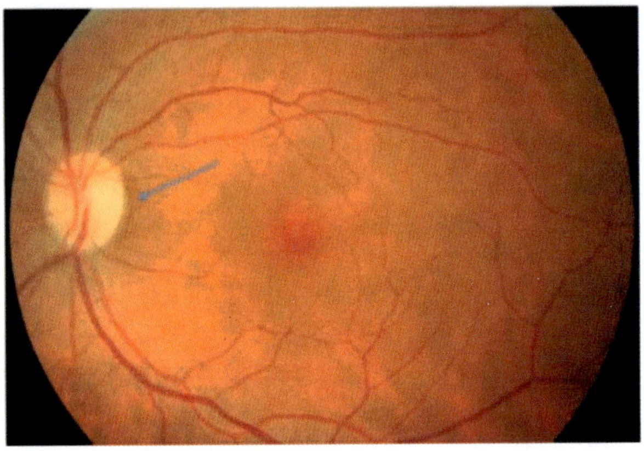

Fig. (9). Normal nerve in one patient photo-type IV in the Fitzpatrick classification. The sharpness of the magnification allows you to carefully observe the characteristics of vessels in the optic nerve; what allowed the objectives of our study, but at these magnifications, the omnipresence of the melanin (blue arrow) in this patient like in the almost six thousand patients studied, powerfully drew our attention.

Optic Nerve in Young (Fig. 10)

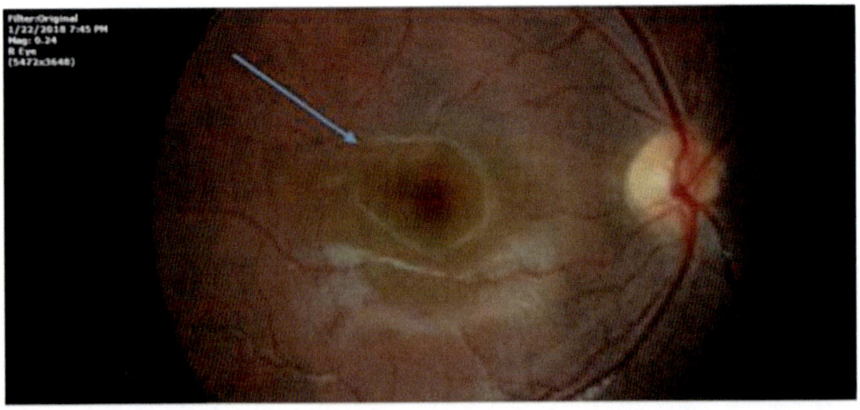

Fig. (10). Tissues in the youth, contain, characteristically a greater amount of water, which is perceived as brightness on the retina. The picture shows the optic nerve, blood vessels that emerge (arteries) and those who penetrate (veins) together with the axons of the ganglion cells layer of the retina. The shape of the macula (blue arrow) is typical of very young patients.

Characteristically the tissues of young patients have a greater amount of water, which can be explained by a higher activity of melanin.

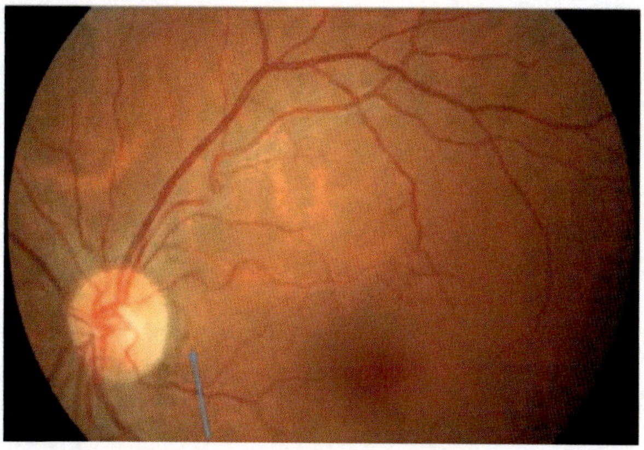

Fig. (11). In the temporal edge, to the right of the picture, there is a delicate brownish line, who seems to replicate the edge of optic nerve and is formed by melanin.

Melanin Nearby the Optic Nerve (Fig. 12)

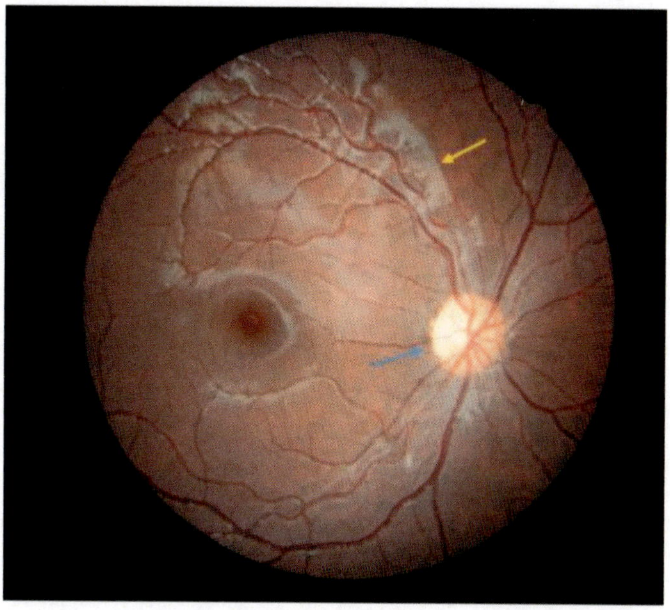

Fig. (12). the insistence of nature to place melanin (blue arrow) in the surroundings of the optic nerve in all patients we studied during the protocol, was the initial observation that aroused our curiosity. Thus, was born our interest in trying to understand the role of melanin in this relatively dark area, given that melanin is considered, for more than one century; a simple sunscreen that protects us from UV rays. The yellow arrow signals the axons that are born from the layer of ganglion cells of the retina and are directed to the optic nerve on its way to the CNS, ending, in its first synapse, in the lateral geniculate body of the thalamus.

Melanin and Macular Diseases (Fig. 13)

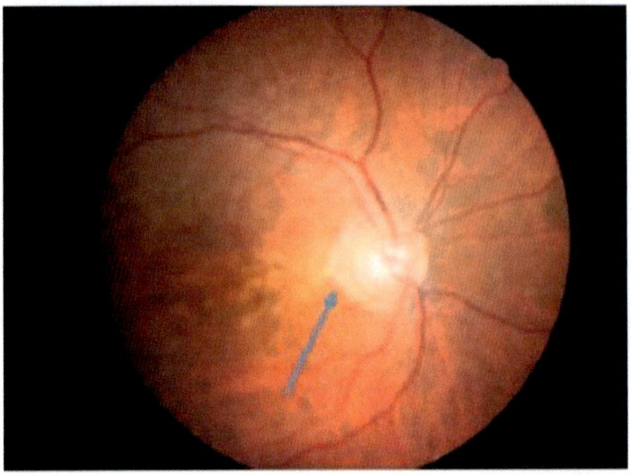

Fig. (13). Melanin invariably accompanies the healthy and diseased eye tissue. When the retina has inflammatory chronic process of any etiology, even purely mechanical; melanin seems to activate and increase synthesis initially, but in long lasting cases, melanin tends to disappear (blue arrow).

Choroidal Nevi (Fig. 14)

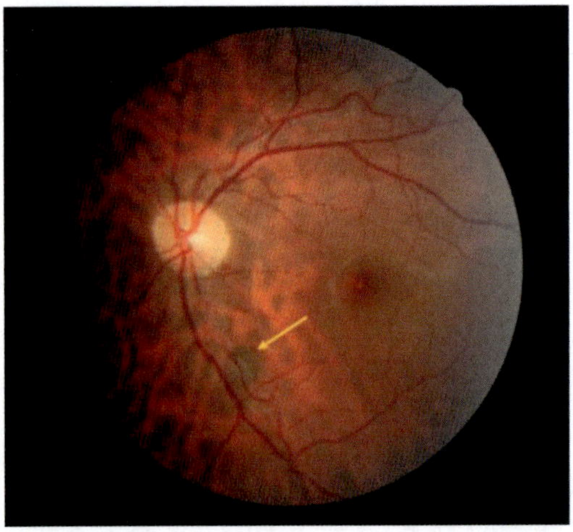

Fig. (14). In some patients, it is possible to find clumps of melanin in the periphery of the retina, which are called choroidal nevi (Yellow arrow). In this photography, a pale area can be seen around of the Nevus, the difference in color is given by a smaller number of blood vessels compared to the normal retinal tissue that surrounds the Nevus. What we call anti angiogenic effect of melanin, is surprisingly consistent. In fact, it was one of the key data that allowed us, twelve years later; to discern the capacity of melanin dissociate the molecule of water, the oxygen levels in the areas of increased pigmentation are significantly higher than the least pigmented areas, and on the other hand, important levels of oxygen have a powerful anti-angiogenic effect.

Pigmentary Retinitis (Fig. 15)

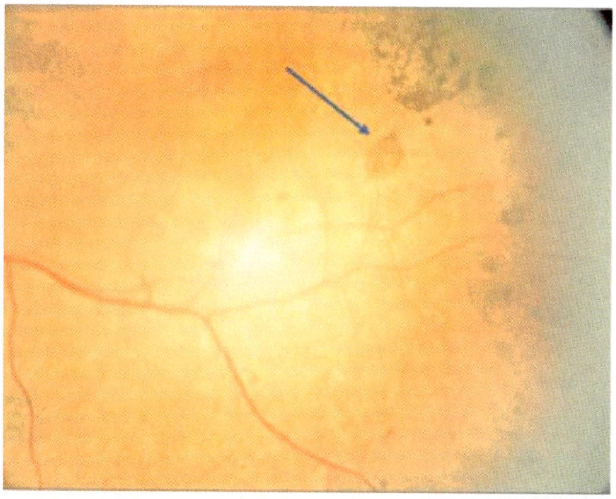

Fig. (15). In retinal degenerations also melanin (blue arrow) makes an appearance. Until now it was thought that first retinal tissue is altered and then melanin simply migrated or stacked, but in our opinion, given the enormous metabolic importance of the chemical energy that melanin produces, is possible to think that first the melanin is altered and then come the alterations of tissues.

Electromagnetic Radiation (Laser) Activates the Synthesis of Melanin (Fig. 16)

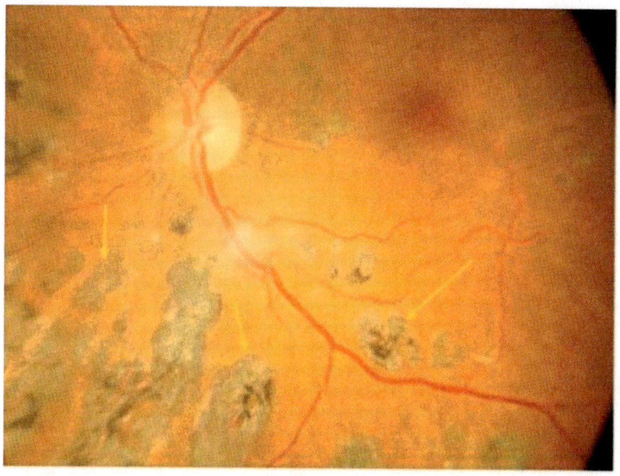

Fig. (16). The picture corresponds to a diabetic patient with scars from Laser photocoagulation (yellow arrows). Melanin appears to be accumulated in the center of the scars of Laser and the vascular changes around them are difficult to analyze for the confounding factors such as the possibility that the difference in the number of vessels would be given by the destruction of tissue caused by the intense applied energy. But in studies reported in the literature, where levels of oxygen measurements were made in the intraoperative surgery of vitreous or retina, detected that the oxygen levels were higher in the pigmented areas than non-pigmented areas, contrary to what was previously thought.

Melanin and Myopia (Fig. 17)

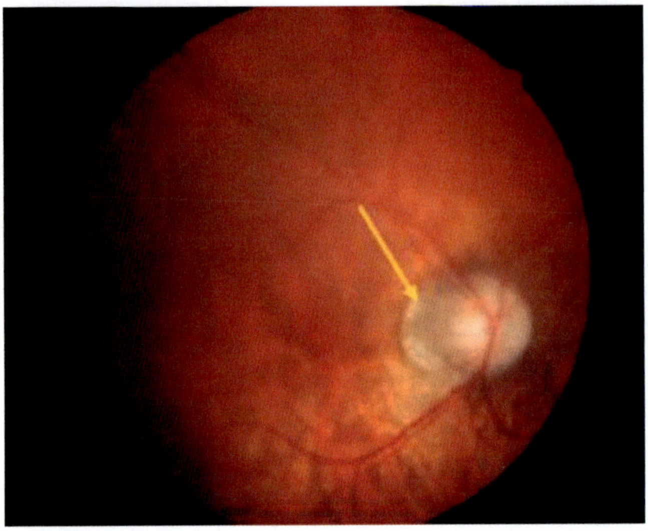

Fig. (17). Melanin always accompanies any eye disease and is also present in the healthy eye. This insistence of nature took us to try to understand why. The picture corresponds to a patient with high axial myopia in where you can see a small optic nerve and it's around an accumulation of melanin (yellow arrow), surrounded in turn by an area of tissue with few blood vessels and some stretch marks also of melanin, which extends to forming a Crescent bordered by melanin and followed by choroid-retinal tissue of normal appearance.

Sometime during the studio which lasted for 12 years (1990-2002), the observation that the blood vessels seemed to respond to the presence of melanin, allowed us to reach the conclusion that a greater amount of melanin, fewer blood vessels and *vice versa*. Sooner or later, we were convinced that melanin had an anti-angiogenic effect very consistent, so much so, that it could not explain by a process mediated by receptors, as the size of the effect is enormous. And found significantly elevated levels of oxygen in pigmented tissues, which in some structures reach a 34% difference, was that we began to look for the source of so much molecular oxygen. The amount of oxygen is so high and so constant, who had to come from a "renewable" source, because if any molecule or molecules of eye tissue to donate it, sooner or later disappear or at least changes noticeably. But any molecule in the eye exhibits a similar behavior. So, the only possibility that remained was that melanin was able to dissociate the water molecule, which is congruent with the fact that more than 90% of the eye is water. The chemical reaction that we identified at the start was as follows:

$$2H_2O \rightarrow 2H_2 + O_2$$

but in laboratory tests, when broke melanin and water, the water did not abate despite the passage of months or years; and that led us to conclude that reaction, in melanin, unlike chlorophyll; It was reversible and could write in the following manner:

$$2H_2O \rightarrow 2H_2 + O_2 \rightarrow 2H_2O + 4e^-$$

Or so:

$$2H_2O \leftrightarrow 2H_2 + O_2 + 4e^-$$

The presence of melanin affects not only the size and number of blood vessels, also has an effect that we call trophic on surrounding tissues, which are observed with greater vitality, thicker, could say that healthier. And this can be explained by the higher levels of chemical energy result of the dissociation of the water molecule. But the energy that is released requires to be transported and molecular hydrogen is the ideal candidate due to hydrogen is the carrier of energy par excellence in the entire universe, thereby; our body cannot be different.

This allows also discard the dogma that the oxygen is essential to generate chemical energy through glucose combustion (oxidation), which was devoid of sense, as the somewhat loud combination that happens between the oxygen and glucose, produces CO (carbon monoxide) mainly and lesser amounts of CO_2 (carbon dioxide), which is a small but significant difference, overall in a biological system.

Melanin releases energy symmetrically, in all directions, by way of increased energy spheres, and given that the melanosomes are predominantly in the perinuclear space, is an ideal place to flood the cell cytoplasm with molecular hydrogen, which is gaseous and spreads following the laws of simple diffusion, by providing energy to the cell nucleus also as this does not have neither mitochondria nor ATP, which solves the riddle about where came the energy that the cell nucleus requires.

Melanin Releases Energy Symmetrically in all Directions (Fig. 18)

Fig. (18). Melanin, represented by circle dark, central; it releases energy, carried by molecular hydrogen (H_2); in an equivalent way to growing spheres, which in turn a small bubbles (gaseous) molecular hydrogen are formed. Each sphere would contain different concentration of molecular hydrogen, they would alternate each other with spheres of re-formed water; and for every two molecules of water re- formed, 4 high energy electrons would be generated which also means energy.

CONCLUSIONS

When a considerable number of atoms is arranged with regularity near, for instance, in a crystal lattice, the terms of the single valence electrons may fuse in common bands. The electrons in this band cease to belong to one or two atoms only, and belong to the entire system. Thus, the absorption and emission of energy will proceed independently at separate places. These energy levels that can be called common exist also in living systems.

The first clue of the existence of processes such common energy levels was given by the study of photosynthesis in plants. Biological molecules, in spite to be built up of a substantial number of atoms, are closely packed with great uniformity; with noticeable consistency, in analogous conditions to those in crystal prevail. That 2500 chlorophyll molecules form one functional unit, and the electrons, raised to a higher energy level by the absorbed light, can move and transport their energy easily through the complex system of chlorophyll molecules, it is not by chance. There is something behind these ordered arrangements. Melanin can be the explanation. It is the most stable molecule man known, and its output of energy is highly consistent, even day and night. Therefore, the chemical reactions that melanin energy impels, are also very ordered, and this energy can drive the behavior of atoms in a unique way, forming molecules relatively simple, but with growing complexity; meanwhile are maintained under the influence of the constant and highly ordered output of molecular hydrogen and high energy electrons.

Thereby a vast number of molecules may join to form such as energy continuum form, and along with energy, excited electrons may travel significant distances at light speed, dissipating energy at the same time.

Prof. Szent-Gyorgyi idea of common levels of energy was important, however, he cannot achieve a coherent explanation due to accept the wrong concept of ATP as source of energy and never mentioned melanin in his works.

The unsuspected intrinsic property of the melanin dissociating and reforming the water molecule, therefore, transduces light into chemical energy [16]; constitutes a turning point in relation to the long-sought answers as eukaryote cell coupled energy with cellular processes. The concept of common energy levels stating Prof. Szent-Gyorgyi are compatible with the way in which melanin releases chemical energy in the form of molecular hydrogen and high-energy electrons.

Furthermore, the detailed experiments of Prof. Szent-Gyorgyi in muscle tissues yield results that are very compatible with the existence of the chemical energy that comes from the melanin, then with the current theory that the muscular energy comes from ATP. Break into a thousand pieces an old paradigm, the clear demonstration of the Prof. Szent-Gyorgyi that energy levels are higher in a muscle relaxed compared to a contracted muscle; therefore, contraction represents a significant decrease in energy levels. The problem of how the energy liberated by a ATP molecule can be communicated to a substantial number of myosin molecules is resolved if we take in account the common energy levels and the fact that ATP absorbs energy of the system and is transformed in ADP. Once energy levels are turned down, the myosin tends to contract, this is: its energy level now is diminished; but common energy level (energy that comes from melanin) is maintained because the adenosine triphosphatase activity is bound up with a very small fraction of myosin molecules but not has effect on melanin activity, thereby the decrease in chemical energy is located and sized precisely.

In any system, biological or otherwise, it is more efficient and easier to lower levels of energy than lift them, this might explain why catabolic processes prevail over anabolic ones.

Now we can understand that numerous diseases accompanied with pain tense muscles, constrained by a greater or lesser degree simply means low levels of the basal chemical energy levels. In the background, a spastic muscle has insufficient chemical energy levels within the muscular tissue; therefore, the muscle tends to shrink. The night cramps in the lower limbs we see for example in diabetics reflect a low sharp level energy, so the muscle shrinks abnormally and is accompanied by pain, but in diabetes mellitus, common energy levels are usually impaired by several reasons, so the muscle takes much time relax, since to do so it requires to raise your energy level starting from the chemical energy that emanates from the melanin.

The finding, that muscle is relaxed when the system is at its maximum power level, links of muscle spasticity with metabolism in poor oxygen supply, changes temperature is very interesting for management of myopathies, muscle pain, fatigue, trigger points, neuromuscular and neurodegenerative conditions. Considering impact of the light on muscle work, energy cumulating may contribute for supporting preventive medicine for wellbeing. In fact, a usual treatment in muscle disorders is the application of long wavelength light, which is used in rehabilitation medicine. Muscle is very appropriate tissue to study energy balance, and start point to implement to the clinic. Melanin can transform light visible and invisible into chemical energy through the dissociation of water molecule, as chlorophyll in plants; however, in the leave the process is irreversible, and the heme group only works absorbing extremes of visible light. It is conceivable that similar process occurs in myoglobin and in other compounds that also have heme group as cytochrome P 450. The prevention of diseases is an imperative, given that morbidity and mortality from several epidemiologically significant diseases seem to be increasing. You cannot prevent or much less cure what you don't understand.

Go back to the concept of Prof. Szent-Gyorgyi in the context of the amazing intrinsic property of melanin transform (visible and invisible) light energy into chemical energy through the water molecule dissociation, as chlorophyll in plants, represents a 180° turn in the current cell biology. While faster to break the old paradigm of glucose as energy source, it will be better for the advancement of knowledge, and patients will benefit in a more comprehensive, faster way. We can prevent, predict and customize the medicine in a much more accurate and efficient way. Fortunately, some publications began to appear in that sense [17].

HUMAN AND ANIMAL RIGHTS

No humans were used in this research. All animal research procedures followed were in accordance with the standards set forth in the eighth edition of Guide for the Care and Use of Laboratory Animals published by the National Academy of

Sciences, The National Academies Press, Washington, D.C.

CONFLICT OF INTEREST

The author (editor) declares no conflict of interest, financial or otherwise.

ACKNOWLEDGEMENTS

This work was supported by Human Photosynthesis® Research Center

REFERENCES

[1] Wurtman, R.J. The effects of light on the human body. *Sci. Am.,* **1975**, *233*(1), 69-77. [http://dx.doi.org/10.1038/scientificamerican0775-68] [PMID: 1145170]

[2] Robin, A. *Biological perspectives in Human Pigmentation*; Cambridge University Press: New York, **1991**, p. 89. [http://dx.doi.org/10.1017/CBO9780511600463]

[3] Szent-Györgyi, A. Towards a new biochemistry? *Science,* **1941**, *93*(2426), 609-611. [http://dx.doi.org/10.1126/science.93.2426.609] [PMID: 17841996]

[4] Livingston, R.; Franck, J. Remarks on intra- and intermolecular migration of excitation of energy. *Rev. Mod. Phys.,* **1949**, *21*, 505-509. [http://dx.doi.org/10.1103/RevModPhys.21.505]

[5] Solis-Herrera, A. Photo electrochemical method of separating water into Hydrogen and Oxygen, using melanins or the analogues, precursors or derivatives thereof as the central electrolyzing element. United States Patent No.: US 8 455,145 B2, Jun. 4, 2013.

[6] Herrera, A.S.; Esparza, Del C.A.M.; Md Ashraf, G.; Zamyatnin, A.A.; Aliev, G. Beyond Mitochondria, what would be the energy source for the cell? *CNS Agents Med. Chem.,* **2015**, *15*, 32-41.

[7] Westheimer, F.H. Why nature choose phosphates. *Sci. New Ser.,* **1987**, *235*(4793), 1173-1178.

[8] Szent-Gyorgyi, A. Free-energy relations and contraction of actomyosin In: *Muscular Contraction*; Acad. Press: New York, **1947**.

[9] Rayment, Ivan The structural basis of the myosin ATPase activity. *J. Biol. Chem.,* **1996**, *271*(27), 15850-15853.

[10] Szent-Gyorgyi, A Ascorbic Acid in Aquatic Organisms: status and perspectives.*Nature of Life*; CRC Press LLC, **2001**.

[11] Szent-Gyoergyi, A.; Kaminer, B. Ka miner, B. Meting and Met actomyosin. *Proc. Natl. Acad. Sci. USA,* **1963**, *50*, 1033-1036. [PMID: 14096175]

[12] Szent-Gyorgyi, A.G.; Kaminer, B. Meting and Met actomyosin. *Proc. Natl. Acad. Sci. USA,* **1963**, *50*, 1033-1036.

[13] Szent-Gyorgyi, A.G. Bioenergetics. *Science,* **1956**, *124*(3227).

[14] Szent-Gyorgyi, A.G.; Borbiro, M. Depolymerization of light meromyosin by urea. *Arch. Biochem. Biophys.,* **1956**, *60*, 180. [PMID: 13283602]

[15] Emerson, R. Arnold, W. Jour. A separation of the reactions in photosynthesis by means of intermittent light. *J. Gen. Physiol.*, **1931**. *16*: 191.

[16] Suchkov, S.; Solís-Herrera, A. The role of Human Photosynthesis® in predictive, preventive and personalized medicine. *EPMA J.*, **2014**, *5*(Supp.l 1): A146.

[17] Brenner, S. Parkinson's disease may be due to failure of melanin in the Substantia Nigra to produce molecular hydrogen from dissociation of water, to protect the brain from oxidative stress. *Med. Hypotheses,* **2014**, *82*(4), 503.
[http://dx.doi.org/10.1016/j.mehy.2014.01.013] [PMID: 24529916]

SUBJECT INDEX

A

Actin and myosin 259, 261, 264, 266, 267, 268, 269, 271
Activation energy 104, 105, 122, 149, 174, 186, 195, 258
Actomyosin 259, 266, 269, 270, 271, 272, 278, 283
Adenine nucleotide translocase (ANT) 245
Adenosine levels 135, 136
 intracellular 136
ADP and chemical energy 269
ADP energy 269, 277, 283
Age-related macular disease 206, 212
Albinism 82, 84, 85
Albinos 81, 82, 248
Amino acids 30, 105, 121, 125, 178, 186, 206, 208
Amyloid beta 202, 203
Angiogenesis 203, 204, 205, 209, 210, 211, 220, 225
Anti-VEGF antibody 220, 223, 226
ATP, hydrolyzed 269, 277
ATP binding 262, 263, 264, 267
ATP concentration 180, 265, 266
ATP concentration of muscle 261
ATP energy 269, 284
ATP hydrolysis 105, 144, 180, 238, 263, 264, 267, 268, 269, 271, 276, 278, 279, 280, 283
ATP hydrolysis in myosin 262, 263
ATP molecules 105, 143, 144, 158, 187, 238, 240, 247, 262, 268, 271, 278, 281, 292
Atrophy, cortical 190, 193, 195

B

Basal chemical energy 283
Bioenergetics and heart muscle 275
Biological molecules 80, 237, 291
Biological realms 7, 8, 26
Biology of saccharides 118, 141
Biosynthetic reactions 104, 180
Blood 51, 92, 97, 98, 99, 119, 122, 125, 134, 142, 144, 157, 159, 163, 165, 167, 168, 169, 170, 171, 172, 173, 174, 175, 176, 233, 239, 242, 262, 269
 arterial 97, 170, 176
 venous 98, 169
Blood cells, red 168, 174, 238, 239, 250
Blood circulation 169, 173, 270
Blood flow 97, 98, 269
Blood oxygen tension 172
Blood plasma 145, 157, 170, 223
Body cells 198
Body fluids 157
Body temperature 172, 233
Brain tissues 122, 124, 125, 193, 195, 196, 204, 205
 intact 124, 125

C

Cancer cell 239, 240, 241
Cancer cells and embryonic tissues 239
Carbohydrates 30, 119, 121, 131, 233, 237, 256, 280
Carbon dioxide 102, 157, 158, 159, 161, 163, 164, 165, 169, 170, 171, 172, 237, 290
Cells 68, 69, 70, 72, 73, 74, 75, 76, 81, 82, 87, 89, 91, 92, 100, 103, 104, 105, 106, 109, 111, 112, 118, 119, 120, 121, 122, 123, 124, 125, 126, 128, 129, 130, 131, 132, 133, 134, 135, 136, 141, 142, 143, 145, 146, 148, 149, 150, 154, 156, 157, 158, 163, 164, 165, 168, 169, 170, 171, 172, 177, 178, 179, 180, 182, 183, 184, 186, 187, 193, 194, 195, 196, 197, 198, 199, 205, 206, 207, 208, 223, 224, 230, 233, 234, 236, 237, 238, 239, 242, 244, 245, 246, 248, 249, 250, 254, 255, 256, 257, 258, 266, 268, 278, 280, 282, 284, 286, 292
 brain 124, 125

eukaryote 193, 198, 250, 254, 292
ganglion 223, 286
human 143, 194
living 68, 73, 74, 109, 130, 180, 256
melanin-containing 92
mesangial 134
nerve 154, 248
neuron 195, 197, 199, 233
single 195, 242, 246
yeast 238
Cerebral artery 151, 153
Cerebral vascular events 150, 151, 152, 153, 198
Chemical energy 52, 121, 133, 103, 258, 267, 277
common 277
consistent 267
fundamental 121
ordered 133
producing free 103
releasing 258
supplying 52
Chemical energy availability, higher 263
Chemical energy compounds 158
Chemical energy levels 122, 263, 264, 267, 271, 272, 273, 276, 283
Chemical energy life 256
Chemical energy light energy 54
Chemical energy result 290
Chemical energy source 141, 262
fundamental 141
Chemical energy variations 261
Chlorocruorins 156, 157, 168
Chlorophyll in plants 49, 61, 80, 138, 154, 156, 187, 189, 193, 199, 230, 271, 293
Chlorophyll molecules 173, 256, 274, 279, 291
Chlorophyll of plants 64, 101, 102
Choroid layer 4, 6, 103, 246
Chromatin 178, 186
Chromosomes 177, 186
Concentration of melanin 9, 10, 28, 77, 98
Connective tissue growth factor (CTGF) 225
Contaminated water 106, 250
Contractile matter 259, 260, 261, 265, 266, 270, 283
CSF production 144, 145
Cytoplasm 121, 131, 142, 147, 179, 180, 183, 184, 187, 195, 258
Cytosol 82, 126, 128, 129, 130, 146, 147, 178, 238, 245, 249

D

Dark melanin molecule 79
Diatomic hydrogen 99, 100, 128, 170, 207, 242, 244, 249, 258, 270, 280, 281
Discovery of human photosynthesis 111, 112
Disorder 70, 71, 74, 110, 111, 120, 189, 263
cardiovascular 110
mental 189
Dissociate and re-form 87, 91, 96, 98, 99, 102, 111, 156
Dissociated water molecule 102
Dissociate water molecule 120, 277
Distribution of energy 63, 118, 132, 152, 154, 169, 175, 196, 250, 264, 267, 271, 273, 274, 276, 280
Drug-melanin interactions 1, 80

E

Edema 107, 146, 185, 198, 202, 211, 212, 217, 226
Electromagnetic radiations 91, 183, 255, 283, 288
Electrons, high-energy 99, 103, 128, 132, 133, 148, 270, 292
Electroretinogram 221
Embryonic tissues 239, 241
Endothelial cells 96, 134, 204, 223, 224, 226
vascular 96, 204, 223
Energy 72, 81, 109, 121, 122, 142, 146, 149, 179, 182, 183, 184, 189, 193, 194, 195, 197, 208, 235, 239, 244, 249, 256, 257, 263, 268, 277, 278, 279, 283, 291
absorbed 146, 249, 268
absorption of 263, 279
availability of 72, 195
available 72, 122, 194, 208, 257

basic 277, 283
codified 72
electrical 81
emission of 279, 291
excess 256, 279
excess of 278
highest 183, 184
luminous 149
nucleus 179
photonic 109, 182, 235
primary source of 121, 122, 142
real source of 189, 193, 208, 239, 244, 257
spheres of 277
tissue 197
Energy carrier 129, 146, 148, 170, 184, 194, 207, 249, 256, 290
Energy conservation 72
Energy conversion 70, 262
Energy Cost 105
Energy currency 237, 265
Energy dissipation 2, 71
Energy dissociates water 181
Energy electrons, flow of high 129, 148, 244
Energy expenditure 69, 125, 126, 128, 130, 156, 163, 179, 186
chemical 125, 128, 130
Energy failure 125, 195
Energy function, unlikely 149
Energy generation 70, 130, 264
Energy levels 111, 122, 143, 183, 260, 263, 266, 267, 268, 271, 272, 278, 279, 280, 283, 284, 291, 292, 293
basal chemical 263, 283, 293
body modulate 283
common 268, 292
complete system 284
higher 122, 183, 267, 279, 283, 291
muscular chemical 272
resting muscle 268
Energy losses 72
Energy melanin 100
Energy metabolism 123, 136, 164
Energy provider, main 133, 142
Energy requirements 125, 131, 149, 178, 246, 247
Energy supply 146, 198
Energy transduction steps 263, 270
Entropy 63, 70, 71, 72, 73, 74, 106, 111, 267, 272
Entropy reduction 69, 73, 74
Enzyme carbonic anhydrase 169, 170, 172, 174
Epithelial cells 203, 239
Erythrocytes 168, 170, 174, 239
Eukaryotic cells, human 208, 233
Eumelanin 81, 86, 88, 246, 247, 248, 249
Events, energy transduction 263, 267
Evolution and entropy 71, 72
Evolutionary cell biology in terms of energy conservation 72
Exchange of energy 149, 186, 220, 267, 277
Extracellular deposits 203, 205
Eye lens 238, 239, 250

F

Factors, vascular endothelial growth 204, 224
Free chemical energy 91, 96, 100, 101, 102, 104, 105, 106, 109, 111, 123, 233, 234, 235, 236, 237, 241, 244
Free energy 75, 105, 112, 180
chemical 75, 105, 112
Function 2, 4, 6, 7, 31, 44, 49, 50, 52, 53, 55, 56, 57, 62, 75, 80, 81, 82, 84, 85, 91, 92, 100, 105, 109, 112, 118, 119, 121, 122, 124, 127, 130, 131, 132, 133, 141, 142, 144, 145, 147, 154, 158, 170, 173, 174, 175, 177, 181, 183, 186, 195, 199, 208, 217, 220, 224, 227, 230, 233, 240, 244, 246, 249, 257, 267, 278
energy-carrier 147
main 2, 7, 56, 80, 127, 133, 144, 181
retinal 224
Functional units 92, 134, 260, 261, 279, 291
Fundamental molecule of life and evolution 75

G

Gas exchange 157, 164
Gluconeogenesis 105
Glucose levels 121, 122, 123, 142, 274

elevated blood 122, 123
Glucose metabolism 104, 121, 122, 123, 124, 142, 234
critical role of 121, 142
Glucose molecule 118, 256
Glycans 118, 119, 120, 121, 123, 124, 125, 126, 127, 128, 130, 131, 141, 142
Glycobiology 118, 119, 130, 131, 141
Glycoconjugates 125, 126, 127, 128, 129, 130, 131, 147
Glycolipids 120, 126, 127, 129
Glycolysis 104, 105
Glycoproteins 119, 120, 126, 127, 129, 130
Glycosylation 130, 131
Growing spheres of energy 148, 196, 246

H

Heat production 89, 90
Heme group of myoglobin 265, 284
Heme groups 157, 166, 171, 174, 254, 274, 275, 280, 284, 293
Hemerythrin 156, 157, 168
Hemocyanin 156, 157, 167, 168, 174
Hemoglobin 77, 127, 156, 157, 159, 166, 167, 168, 169, 170, 172, 174, 254, 275, 280, 284
Hemoglobin heme groups 157, 166, 168, 171
Hemoglobin molecule 166, 168, 170
Hemorrhage 146, 185, 202, 211, 212, 214, 215, 216
Heterogeneity 87, 130
Hexoses 125, 126
High level brain function 121, 142
Hippocampus 107, 135, 136, 190, 193
Hydrogen and oxygen 64, 102, 128, 194
Hyperglycemia 122, 124

I

Intraocular injection 220, 223, 226, 228
Intrinsic capacity of melanin to split and re-form 88, 90
Intrinsic property of melanin to dissociate and re-form 87, 91, 96, 98, 99, 111
Intrinsic property of melanin to split and re-form 88, 109, 234, 237, 245, 246

K

Kleiber's law 89, 90, 99, 101

L

Lactic acid 239, 262, 281, 282, 283, 284
Law of increasing entropy 71, 72
Levels 63, 111, 135, 136, 138, 162, 195, 264, 267, 270, 271, 272
basal 135, 264, 271, 272
energetic 267
higher oxygen tissue 138
intracellular 135, 136
molecular 63, 111, 162, 195, 270
Light energy 52, 58, 77, 78, 89, 96, 102, 127, 133, 138, 149, 154, 165, 170, 171, 172, 175, 184, 256, 257, 266, 293
melanin transform 154, 165
Light-melanin-water 177
Light of human PhotosynthesisTM 238, 241, 244
Lignin 1, 8, 58, 77, 78
Locus coeruleus (LC) 195, 199, 246

M

Magnetic resonance imaging (MRI) 190, 193
Melanin of 7, 8, 18, 44, 45, 50, 57, 62, 91, 93, 96, 229
abundant 44, 55
omnipresence of 7, 8, 18, 50, 57, 62, 91, 93, 96
oxygen diatomic levels nearest 229
Melanin and myoglobin 270, 271, 273, 277, 278
Melanin content 4, 137, 144, 243
high 4, 144
Melanin dissociates 64, 236, 265, 287
Melanin energy impels 292
Melanin Functions 2, 6, 80, 91

Melanin granules 247, 249
Melanin molecule, single 194
Melanin Photo-System 183, 185
Melanin pigments 75, 80, 82, 92
Melanin release energy 194
Melanin research 80, 86
Melanin skin 50
Melanin synthesis 1, 84
Melanin transform 49, 293
Melanin works 245, 249
Melanocytes 84, 176, 246
Melanogenesis 82, 86
Melanoma 137
Melanophores 1, 246
Melanosomes 82, 87, 100, 128, 129, 146, 148, 183, 243, 246, 248, 290
Melanosomes form 129, 148
Meromyosins 275, 276
Metabolic activities 89, 92, 124, 134, 148, 224
Metabolic rate 89, 90, 100, 131, 163
Midbrain 4, 248, 277
Mitochondria functions 244
Mitochondrial DNA 238, 249
Mitochondria number 137, 243, 244
Molecular biology 119, 154
Molecular hydrogen 123, 129, 133, 147, 168, 171, 182, 184, 186, 194, 195, 196, 198, 199, 235, 249, 254, 262, 264, 265, 266, 268, 269, 271, 272, 276, 277, 280, 283, 284, 290, 291, 292
Molecular masses 157
Molecular material 74
Molecules 149, 166, 168, 180, 207, 237, 242, 249, 275, 278, 291
  bio-energetic 275
  bio-energetics 278
  large 166, 168, 180
  nutrient 237
  stable 149, 168, 207, 242, 249, 291
  water-soluble 180
Molecule water 98
Monoclonal antibodies 203, 220, 223, 225, 227
Multifactorial diseases 204
Muscle 52, 83, 84, 104, 122, 142, 164, 167, 168, 171, 173, 206, 208, 233, 257, 258, 259, 260, 261, 262, 264, 265, 266, 267, 268, 269, 270, 271, 272, 273, 274, 275, 279, 280, 283, 284, 292, 293
  ciliary 83, 84, 270
  contracted 259, 266, 267, 292
  turkey breast 283, 284
Muscle contraction 254, 259, 261, 264, 265, 266, 269, 272, 283
Muscle disorders 254, 293
Muscle energy 283
Muscle fibers 83, 84, 259, 273
Muscle membrane 271, 272
Muscle relaxation 253, 261, 264, 265, 268, 269
Muscle research 254
Muscle tissue 266, 272, 292
Myoglobin 167, 174, 254, 265, 266, 269, 270, 271, 273, 277, 278, 280, 284, 293
Myosin 254, 259, 260, 261, 262, 263, 264, 265, 266, 267, 268, 269, 270, 271, 275, 276, 279, 292
Myosin molecules 260, 279, 292

N

Nebulin 259, 260, 261, 267
Neuromelanin 195, 199, 247, 248
Neuro-melanin 59, 87, 196
Nitrogen 86, 150, 158, 159, 198, 237, 255
Nuclear envelope 178, 179, 180
Nuclear pores 126, 178, 179
Nucleon-plasm 179, 180
Nucleoside 134, 135, 136
Nucleotidase 135
Nucleotides, low-molecular-weight sugar 128, 146
Nucleus energy requirements 129
Nucleus of eukaryotic cells 146, 181

O

Oligosaccharides 121, 125, 126, 127, 142
Optic nerve 3, 4, 5, 49, 61, 64, 84, 85, 92, 93, 96, 206, 285, 286, 289
Organic matter 235, 254, 255, 257

Organic molecules 63, 64, 236
Oxidative mechanisms 283, 284
Oxidative phosphorylation 240, 242, 256, 257
Oxygen 2, 97, 99, 123, 156, 157, 158, 159, 160, 161, 162, 163, 165, 168, 169, 171, 172, 173, 174, 175, 182, 194, 207, 229, 237, 238, 239, 242, 289
  absence of 238, 239
  absorption coefficient of 159
  absorption of 159, 172
  amount of 165, 172, 173, 289
  concentration of 158, 160, 161
  diatomic 99, 194, 207, 229
  diffusion of 163, 175
  dissolved 159, 242
  entrance of 163, 169
  high pressure of 97, 162
  molecular 2, 99, 157, 168, 182, 242, 289
  molecule of 168, 171
  role of 123, 237
  solubility of 169, 242
  source of 159, 160
  tensions of 161, 171, 175
  transport of 156, 174, 175
Oxygenation 173, 174
Oxygen consumption 89, 241
Oxygen content 97, 98
  molecular 98
Oxygen diatomic 102, 168, 181
Oxygen diffuses 160, 163
Oxygen levels 89, 150, 169, 172, 174, 241, 287, 288
  low 150, 241
  molecular 169
Oxygen molecules 158, 159, 171
Oxygen species, active 81
Oxygen supply 163, 164, 171, 282, 284, 293
Oxygen tension 160, 172, 173, 174
  alveolar 172
  venous 171, 173
Oxygen uptake 164, 165, 171
Oxyhemoglobin 172, 173

P

Partial pressure of oxygen 158, 159, 161, 169
Pheomelanin 31, 86, 247
Phosphate groups 122, 238, 240, 244
Photoreceptor layer 96, 97, 98, 103
Photosynthesis 68, 101, 102, 103, 106, 109, 111, 118, 131, 141, 146, 156, 161, 177, 205, 208, 234, 257, 279, 291
Photo system 181, 182, 185
Photo-system dissociates water 103
Physical forces 69, 73
Plants melanins 247
Plumage 39, 40, 43, 47
Porphyrin 166, 167
Preventive medicine 101, 110, 112, 293
Producing hydrogen 101, 102
Protein molecules 120, 279
Protein synthesis 105, 186, 224
Pyruvate 123, 124

R

Re-formed water molecule 89
Reformed water molecules 128, 194
Repulsive forces 260, 264, 266, 269, 271, 280
Respiratory surfaces 160, 163, 164, 165, 171, 175
Retinal cells 206
Retinal tissues 104, 206, 214
Rhodopsin 91, 92
Rigor mortis 261, 262, 264, 265, 269

S

Sarcomere 259, 260
Schizophrenia 189, 190, 191, 192, 193, 196
Sclera 5, 6, 229
Senile plaques 202, 204
Skin, dark 55, 81
Skin cancer 2, 7, 81
Skin melanin 2

Skin photo-type 247
Solar filters 1, 4, 12, 43, 57, 62, 138
Solis-Herrera cycle 194, 195, 196, 197, 208
Source 37, 54, 75, 80, 104, 109, 111, 118, 123, 125, 131, 135, 141, 142, 143, 150, 162, 171, 187, 191, 193, 194, 195, 197, 198, 206, 208, 233, 237, 242, 255, 257, 262, 282, 289
  approach energy 37, 54
  primary energy 131, 142
Source of biomass 104, 109, 111, 118, 123, 125, 141, 193, 206, 208, 233
Space 100, 128, 134, 135, 136, 148, 178, 180, 242, 243, 290
  extracellular 134, 135, 136
  perinuclear 100, 128, 148, 178, 180, 242, 243, 290
Species of reptiles 13, 14
System 70, 71, 73, 74, 256
  chaotic 73, 74
  enzyme 256
  thermodynamic 70, 71

T

Tension differences 161
Thermodynamic equilibrium 70, 71, 73
Thermodynamics 69, 70, 72, 74, 106
Tissue energy levels 197
Tissues 1, 3, 6, 23, 24, 52, 58, 63, 69, 73, 75, 85, 90, 92, 93, 96, 98, 99, 104, 122, 123, 125, 131, 134, 135, 136, 142, 144, 146, 150, 153, 154, 156, 157, 163, 165, 169, 171, 173, 174, 189, 195, 197, 198, 204, 205, 206, 208, 211, 214, 216, 221, 225, 227, 238, 239, 240, 241, 243, 249, 250, 254, 262, 267, 270, 271, 284, 285, 286, 288, 289
  edematous 216
  nervous 154, 171, 189
  neuronal 204, 206, 208
  peripheral 122, 123, 142
  pigmented 75, 289
  plexus 144
Transform light energy 49, 75, 81, 90, 96, 102, 118, 123, 128, 132, 156, 168, 173, 189, 230, 254
Transport energy 141, 206, 208, 279
Transport of gases 163, 164
Tropomyosin 276, 278
Tyrosinase 1, 86
Tyrosine 83, 86, 247, 248

U

Unexpected role of melanin and myosin 268
Unsuspected bioenergetic molecule 284
Uveal tract 4, 5, 6

V

Vascular endothelial growth factor (VEGF) 204, 205, 220, 222, 223, 224, 225, 227

W

Water dissociation reaction 145, 146
Water droplets 59, 60
Water energy 265
Water molecule breaks 205, 233
Water molecule dissociates 65
Water molecule dissociation 127, 150, 165, 169, 171, 175, 193, 221, 293
Water reformation 194, 195

Made in United States
Orlando, FL
04 November 2026

90199927R00186